ENCYCLOPAEDIA OF CLASSIFICATION OF FISH

ENCYCLOPAEDIA OF
CLASSIFICATION OF FISH
Vol. 2

A.M. Bagulia

ANMOL PUBLICATIONS PVT. LTD.
NEW DELHI - 110 002 (INDIA)

ANMOL PUBLICATIONS PVT. LTD.
H.O.: 4374/4B, Ansari Road, Darya Ganj,
New Delhi-110 002 (India)
Ph.: 23278000, 23261597
B.O.: No. 1015, Ist Main Road, BSK IIIrd Stage
IIIrd Phase, IIIrd Block,
Bangalore - 560 085 (India)
Visit us at: www.anmolpublications.com

Encyclopaedia of Classification of Fish

First Published, 2008
ISBN 978-81-261-3530-1 (Set)

PRINTED IN INDIA

Printed at Mehra Offset Press, Delhi.

Contents

Preface

Fish are classified in several classes by which the study of fish becomes easier. Fish having similar properties are contained in the same group. Various classes of fish are Thelodonti, Anaspida, Cephalaspidomorphi, Galeaspida, Pituriaspida, Osteostracy, Gnathostomata, Placodermi. Jawless fish are contained under the group Thelodonti, which are very similar to Heterostraci, and are not armoured.

The Pituriaspida are a small group of armoured jawless fish with tremendous nose-like rostrums, which live in the marine, deltaic environments of Middle Devonian Australia. They are known only by two species, *Pituriaspis doylei* and *Neeyambaspis enigmatica,* found in a single sandstone location of the Georgina Basin, in Western Queensland, Australia. The Osteostraci is a group of bony-armoured jawless fish, which lived in North America, Europe and Russia from the Middle Silurian to Late Devonian Period. The Devonian species were among the most advanced of all known Agnathans. This was due to development of paired fins, and their complicated cranial anatomy. However, the osteostracans were more related to other Aghathans then to jawed ventebrates as their inner ears were formed by two pairs of semi-circular canals.

Gnathostomata is the group of vertebrates with jaws. This group is in superclass, including the familiar classes of fish, birds, mammals, and so forth and a sister group of jawless vertebrates Agnatha. The Placodermi are armoured prehistoric fish known from fossils, dating from the Late Silurian to the end of the Devonian Periods. Their head and thorax were covered by articulated armoured plates and the rest of the body was either scaled or naked. Placoderms were among the first of the jawed fish, their jaws likely to have evolved from the first of their gill arches. The first identifiable Placoders evolved in the Late Silurian; they disappeared in the Late Devonian extinctions. Thus, the Fish world is a world in itself.

This eminent work has an exhaustive and exclusive coverage of fish in a simple, but effective style. Let's hope, this work would be capable to prove to be beneficial for all classes of readers.

Editor

1

Breeding

Discus like to choose their own partner. This makes breeding a little more difficult and a lot more expensive. Discus should be kept in groups of at least 4 ideally. If the water parameters are good and they are well fed they will spawn.

Site Choice

The discus will choose a site that is near vertical. They will clean the site. They may even lay their eggs on the bottom or side of the tank.

Laying and Fertilization

The female's egg tube will protrude a few mm when she is laying the eggs. This will make her passes in an upward direction laying the eggs against the surface. The male will follow after her and fertilize the eggs.

Egg Care

The parents will both protect their eggs, chasing away all the fish that come near. They will even eat one at a time so the eggs are not left vulnerable.

They will continually fan their eggs. It is not uncommon for young parents to eat the eggs one at the time. White eggs are infertile.

New Zealand Sand Diver

The New Zealand sand diver, *Tewara cranwellae,* is a sandburrower, the only species in the genus *Tewara,* found all around New Zealand to depths of about 5 m, on sandy or gravelly bottoms. Its length is between 3 and 8 cm.

The New Zealand sand diver is a small cylindrical fish with a pointed snout, a distinctive undershot lower jaw, and small bulbous eyes that swivel independently. They are pale yellow with black markings and the dorsal and anal fins are equal sized.

They spend much of their time buried in sand or gravel with just the tip of the snout and eyes showing, and if they are disturbed they dart to rebury themselves at a new site. They move so fast it is almost impossible follow them, and in consequence are hardly ever seen.

Long-finned Sand Diver

The long-finned sand diver or tommyfish, *Limnichthys polyactis,* is a sandburrower of the genus *Limnichthys,* found all around the North Island of New Zealand to depths of about 5 m, on sandy or gravelly bottoms. Its length is between 3 and 8 cm.

The long-finned sand diver is a small cylindrical fish with a pointed snout, a distinctive undershot lower jaw, and small bulbous eyes that swivel independently. They are pale yellow with black markings. The anal fin is slightly longer than the dorsal fin, and both fins are longer than the very similar New Zealand sand diver. They spend much of their time buried in sand or gravel with just the tip of the snout and eyes showing, and if they are disturbed they dart away to rebury themselves at a new site. They move so fast it is almost impossible follow them, and in consequence are hardly ever seen.

Dogfish

The name dogfish is applied to a number of small sharks found in the northeast Atlantic, Pacific, and Mediterranean, especially those in the three families Scyliorhinidae, Dalatiidae and Squalidae.

Although often used in reference to Scyliorhinus canicula, the name is applied only loosely and does not usually signify a close taxonomic relationship. The Spiny Dogfish (Squalus acanthias) is found on the west coast of the United States including Puget Sound.

They hunt both solitarily and in schools. They eat small fish, squid, and crustaceans, and have extremely strong jaws for crushing the shells. They are considered a nuisance by fishermen because they will latch on to almost anything put in the water. Some fishermen kill them when caught which, along with pollution, has contributed to a sharp decline in population in Puget Sound.

It in now illegal to kill or mutilate them when caught. They are edible, but taste terrible. Care has to be taken when handing because they have 2 poisonous spines at the back of both dorsal fins. The poison is not likely to cause major damage, but the wound can take months to heal.

Dr. Andrew Bailey has done various research in the field of the Spiny Dogfish (*Squalas acanthus*) and was the first to discover about the details of the discharge of the sperm via fine tubules, the vasas efferentia. The spiny dogfish was Dr. Baileys primary field until 2007 as he has now taken up Communications studies.

Squalidae

Squalidae is the family of dogfish sharks. These sharks are characterised by smooth dorsal fin spines; teeth in upper and lower jaws similar in size; caudal peduncle with lateral keels; upper precaudal pit usually present; a caudal fin without subterminal notch; and no anal fin.

Cynodontidae

Cynodontidae fishes (order Characiformes), or dogteeth tetras, are a family of freshwater fishes found in the Neotropics. This group is not very diverse, and includes only five genera and 14 species. Most of what is known about this family is from the members of the subfamily Cynodontinae, which includes the largest species of this family, up to 65 centimetres (26 in). Understanding of the members of subfamily Roestinae are less known, though they only reach up to 20 cm.

Physical Characteristics

They have a very peculiar shape, and their name derives from their elongated and well developed canines which they mainly use to eat other fish. Their pectoral fins are also expanded. The maximum length reached is 65 cm.

Distribution and Habitat

These fish live in mid to surface waters of rivers, lakes, and flooded forests. Most species of this family originate from the Orinoco and Amazon basin. The range reaches as far south as Paraná-Paraguay and Uruguay basins and also includes Venezuela and Colombia. Fossil species are also known from Argentina, where they are not found now.

Relationship to Humans

Hydrolycus are game fish, having been recently added to the International Game Fish Association in the fly and rod class. Cynodontid fish are also sometimes housed in aquaria.

Weather Loach

The Weather loach or Dojo loach (*Misgurnus anguillicaudatus*), is a freshwater fish in the loach family Cobitidae. They are native to Asia but are also popular as an aquarium fish. The name *Weather loach* is shared with several species including the European weather loach (*Misgurnus fossilis*) and the spotted weather loach (*Cobitis taenia*). The name weather loach comes from their ability to detect changes in barometric pressure and react with frantic swimming or standing on end. This is because before a storm the barometric pressure changes, and this is known to make the loach more active.

Like many other loaches, they are slender and eel-like. They can vary in colour from yellow to olive green, to a common light brown or gray with lighter undersides. The mouth of the loach is surrounded by three sets of barbels. It uses them to sift through silt or pebbles to find food. It also uses them to dig under gravel and sand to conceal itself out of nervousness or defence unlike the other loaches who uses the spines beneath the eyes.

They can grow up to a 12 inches (30.5cm) long. The fish are bottom-dwelling scavengers, feeding mainly on organic material such as algae. Weather loaches are omnivorous and may also feed on tubifex worms and other small aquatic organisms. By producing a layer of mucus to keep themselves damp, they can also survive small periods of not being in the water. They are very hardy fish that can live in poor quality water.

In the Aquarium

Weather loaches are active, peaceful, and hardy fish that are sometimes used as starter fish in an aquarium. They can be "friendly" towards humans, allowing physical contact and hand feeding. They have, however, been known to attack very small fish in smaller aquariums or bite humans' fingers. They need to be kept in a minimum of 10 gallons (~30 litres) per fish but a minimum of 4foot by 18inch footprint is needed for weather loaches.

There are other varieties breed from captivity like the gold strain and the peppered strain (not to be confused with the pepper loach)

The loaches will be more active given more space and greater numbers. Solitary weather loaches tend to spend much of their time hiding. They will spend a lot of time hiding or staying still, but should be given a place to stay which will have cover and shade. Tank decorations that they can swim through and driftwood both work great for this.

Due to their ability to survive out of the water, the fish will often jump out of an aquarium. A cover is required to keep them in their tank. Weather loaches enjoy digging and burrowing themselves in the substrate of their tank, so make sure that your substrate is fine enough for them to dig in. If you keep live plants in your tank, they will be uprooted by the loaches, so it is a good idea to weight your plants.

The weather loach is also peculiar in that it will sometimes bury itself in the substrate during times of stress. This often surprises new owners, as the fish will "disappear" shortly after introduction to the tank only to "reappear" later.

Because of their appetite for snails, these loaches can help alleviate snail infestations in tropical fish tanks, though many have reported that while weather loaches do eat snails, they do not eat them at a fast enough rate to deal with an infestation.

The fish prefer a pH of 6.5-8.0 but will tolerate far more acidic conditions even for extended amounts of time with little negative reaction.

This makes the Weather Loach a great choice for first-time aquariums and for those who want a fish tank but do not want the intense, daily attention other fish require.

Dolly Varden trout

The Dolly Varden trout, *Salvelinus malma malma,* is a subspecies of anadromous fish in the salmon family, and is technically a char. Although many of the fish are anadromous, the fish also exists in landlocked waterways in the northwest United States.

Distinguishing Characteristics

The back and sides are olive green or muddy gray, shading to white on the belly. The body has scattered pale yellow or pinkish-yellow spots. There are no black spots or wavy lines on the body or fins. Small red spots are present on the lower sides. These are frequently indistinct. The fins are plain and unmarked except for a few light spots on the base of the caudal fin rays.

Origin of the Name

It appears that the first recorded use of the Dolly Varden name to refer to a species of fish, was to a different species which is similar in appearance (*Salvelinus confluentus*), now commonly known as the "Bull trout."

In his book, Inland Fishes of California, Peter Moyle recounts a letter sent to him on March 24, 1974 from Mrs. Valerie Masson Gomez: "My grandmother's family operated a summer resort at Upper Soda Springs on the Sacramento River just north of the present town of Dunsmuir, California. She lived there all her life and related to us in her later years her story about the naming of the Dolly Varden trout. She said that some fishermen were standing on the lawn at Upper Soda Springs looking at a catch of the large trout from the McCloud River that were called 'calico trout' because of their spotted, colourful markings. They were saying that the trout should have a better name. My grandmother, then a young girl of 15 or 16, had been reading Charles Dickens' Barnaby Rudge in which there appears a character named Dolly Varden; also the vogue in fashion for women at that time (middle 1870s) was called 'Dolly Varden,' a dress of sheer figured muslin worn over a bright-coloured petticoat. My grandmother had just gotten a new dress in that style and the red-spotted trout reminded her of her printed dress. She suggested to the men looking down at the trout, 'Why not call them "Dolly Varden"?' They thought it a very appropriate name and the guests that summer returned to their homes (many in the San Francisco Bay area) calling the trout by this new name. David Starr Jordan, while at Stanford University, included an account of this naming of the Dolly Varden Trout in one of his books."

In 1874, Livingston Stone, a naturalist working for the US government wrote: "Also called at (Upper) Soda Springs the 'Varden' trout. The handsomest trout, and, on the whole, having the most perfect form of all the trout we saw on the McCloud. Also, the only fish that had coloured spots. This one was profusely spotted over most of the body with redish golden spots. The local name at (Upper) Soda Springs is the Dolly Varden". Quoted from: VI. Report of Operations During 1872 at the United States Salmon-Hatching Establishment on the M'Cloud River, and on the California Salmonidae generally; with a list of Specimens Collected. By Livingstone Stone. In: United States Commission of Fish and Fisheries. Part II. Report of the Commissioner for 1872 and 1873. A—Inquiry into the Decrease of the Food Fishes. B—The Propagation of Food-Fishes in the Waters of the United States. With Supplementary Papers. Washington: Government Printing Office, 1874 at pp. 203 - 207.

It is currently unknown whether the name "Dolly Varden" was later applied to this subspecies (*S. m. malma*) because of the similar appearance of the two fish (the two may have been believed to be the same species), or whether the name "Dolly Varden" was given to *S. m. malma* independent of the McCloud River fish.

Ironically, it appears that the "original" Dolly Varden trout (i.e., *S. confluentus*) likely became locally extinct in the McCloud River in the 1970s (although reports continue of its being caught in the McCloud River). Other fish species (typically introduced trout) will out-compete the Dolly Varden/Bull trout, and will actually interbreed with the Dolly Varden/Bull trout, producing sterile hybrids. As a result, an attempt to reintroduce the Dolly Varden/Bull trout to the river where it received its name (the McCloud River) was unsuccessful, and no additional attempts are expected.

Present Day Range

The subspecies *S. m. malma* is found in coastal waters of the North Pacific from Puget Sound to the Alaska Peninsula.

and along the Bering Sea and the Arctic Sea to the Mackenzie River. It is known as 'Belyi golets' in Russian.

The "Dolly Varden" name is also applied to a third subspecies *S. m. miyabei* - which is a landlocked species found on the island of Hokkaido in Japan.

The name has also been applied to still a third species *Salvelinus alpinus*, today more commonly known as Arctic char.

Mahi-mahi

The Mahi-mahi, Coryphaena hippurus, also known as dolphin, common dolphin-fish, dorado maverikos, or lampuka (in Maltese) are surface-dwelling ray-finned fish found in offshore tropical and subtropical waters worldwide. They are one of only two members of the Coryphaenidae family, the other being the Pompano dolphinfish. The name "mahi-mahi" ("strong-strong" in Hawaiian), particularly used on restaurant menus, has been adopted in recent years to avoid confusing these fish with dolphins, which are mammals.

General Chacacteristics

Mahi-mahi have a lifespan of no more than three to four years. Catches average 7 to 13 kg (15 to 28 pounds). They seldom exceed 15 kg (33 pounds), and any Mahi-mahi over 18 kg (39 pounds) is exceptional.

Mahi-mahi have compressed bodies and long dorsal fins extending almost the entire length of their bodies. Their anal fins are sharply concave. They are distinguished by dazzling colours: golden on the sides, bright blues and greens on the sides and back. Mature males also have prominent foreheads protruding well above the body proper. Females have a rounded head.

The males and female have similar shaped bodies except for their heads. Females are usually smaller than the males.

When they are removed from the water, the fish often change between several colours (being this the reason for their name in Spanish Dorado Maverikos), finally fading to a muted yellow-grey upon death.

Mahi Mahi is one of the fastest-growing fish, they are fast swimmers as well, with a top swimming speed of 50 knots. Mahi-mahi spawn in warm ocean currents throughout much of the year; and its young are commonly found in sargassum weed.

Mahi-mahi are carnivorous, feeding on flying fish, crabs, squid, mackerel, and other small fish. They have also been known to eat zooplankton and crustaceans.

Capture

Mahi-mahi are highly sought for game fishing and commercial purposes. Game fishery is popular due to their beauty and fighting ability. Commercially for their flesh which is notable for its flavour and firm texture.

Mahi-mahi have become popular restaurant fare in many areas due to its sweet flesh flavour, sometimes eaten as a substitute for swordfish. It has scales, therefore they are kosher, as well as halal by Shia and Sunni Muslims.

Location

Mahi-mahi can be found in the Caribbean Sea, on the west coast of North and South America, and South East Asia as well as many other places worldwide. The United States and the Caribbean countries are the primary consumers of Mahi-mahi products. Other developed countries in Europe are increasing their consumption every year. Japan is a strong consumer as well. Mahi-mahi play a major role in the novel Life of Pi (they are called dorados).

Dory

The common name *dory* (from the Middle English *dorre,* from the Middle French *doree,* lit. "gilded one") is shared (officially and colloquially) by members of several different families of large-eyed, silvery, deep-bodied, laterally compressed, and roughly discoid marine fish. As well as resembling each other, dories are also similar in habit: most are deep-sea and demersal. Additionally, many species support commercial fisheries and are considered excellent food fish. Most dory families belong to the order Zeiformes, suborder Zeioidei:

- The "true dories", family Zeidae (five species, including the well-known John dory)
- The zeniontids, family Zenionidae or Zeniontidae (seven species)
- The "Australian dories", family Cyttidae (three species all within the genus *Cyttus*)
- The oreos, family Oreosomatidae (ten species)
- The parazen family, Parazenidae (four species, including the rosy dory)

Additionally, several species of spinyfin (family Diretmidae, order Beryciformes) have been given the name *dory* by fishmongers, presumably to make the fish more marketable.

Dory was also the name of a blue tang fish in the Pixar film *Finding Nemo*. Voiced by Ellen DeGeneres, Dory was accompaniment and a friend to Marlin, the father of the title character, Nemo, who got lost in the ocean after a deep sea diver collected him to be added to the diver's personal fish tank. The blue tang—and all tangs—are not included in the Dory fish order Zeiformes, but rather the order Perciformes.

Dottyback

The dottybacks are a family, Pseudochromidae, of fishes in the order Perciformes. Around 100 species belong to this family.

Genera

- *Anisochromis*
- *Assiculoides*
- *Assiculus*
- *Blennodesmus*
- *Chlidichthys*
- *Congrogadus*

- *Cypho*
- *Halidesmus*
- *Halimuraena*
- *Halimuraenoides*
- *Haliophis*
- *Aabracinus*
- *Lubbockichthys*
- *Natalichthys*
- *Ogilbyina*
- *Pectinochromis*
- *Pseudochromis*
- *Pseudoplesiops*
- *Rusichthys*

Dragonet

Dragonets are small perciform marine fish of the diverse family Callionymidae (from the Greek *kallis*, "beautiful" and *onyma*, "name"). Found mainly in the tropical waters of the western Indo-Pacific, the family contains approximately 186 species in 18 genera. The Draconettidae may be considered a sister family, whose members are very much alike though rarely seen. Due to similarities in morphology and behaviour, dragonets are sometimes confused with members of the goby family.

Physical Description

These "little dragons" are generally highly colourful with cryptic patterns. Their bodies are elongate and scaleless; a strong spine guards the preopercle (part of the gill cover), which has been reported to be venomous in some species. All fins are large, showy and elongate; the first high dorsal fin usually has four spines; in males, the first of these spines may be further adorned with filamentous extensions. Dragonets have flattened, triangular heads with large mouths and eyes; their tail fins are fan-shaped and tapered.

The largest species, the longtail dragonet (*Callionymus gardineri*) reaches a length of 30 centimetres. At the other end of the scale, the St. Helena dragonet (*Callionymus sanctaehelenae*) reaches a length of just 2 centimetres. Many species exhibit marked sexual dimorphism: males and females are coloured and patterned differently, and (in addition to the spine filament) males have a much higher dorsal fin. This theme is taken to extremes in the high-finned dragonet (*Synchiropus rameus*).

Behaviour

Dragonets are benthic animals, spending most of their time on or near the bottom. They prefer sandy or rocky substrates, sometimes near reefs. Inhabiting depths down to about 200 metres, dragonets feed mostly on crustaceans, worms and other small invertebrates rooted out from the substrate. The dragonet's large pectoral fins serve as a primary means of propulsion. Males are highly territorial between themselves.

Reproduction

Spawning involves elaborate courtship displays; the males show off their flashy fins and repeatedly open and close their mouths. If the female is interested, pairing occurs and the two fish rise upwards with male supporting the female on his pectoral fins. Eggs and sperm are released in midwater, where fertilization takes place. The buoyant eggs subsequently become part of the plankton, drifting with the currents until hatching.

In Aquaria

The most common commercially available dragonets are the mandarin dragonet, the psychedelic mandarin dragonet and the ocellated dragonet, or scooter blenny. They are considered to be reef safe and do best in reef aquariums of 55 gallons or larger with large amounts of live rock.

There are many reports of spawning in captivity, but it is difficult to find dragonets that have been born and bred in captivity because they require a considerable area of sand or rock to find enough food.

Although their bright colours and showy fins make them a popular choice for the aquarium, most dragonets are picky eaters and will only accept live food, making them difficult to keep in captivity.

They do not readily accept prepared foods and often starve soon after purchase. As stated earlier, larger tanks are desired because they can support a larger population of the copepods and amphipods which make up the bulk of their diet in the wild. Some success has been had feeding brine or mysis shrimps.

Dragonfish

The common name dragonfish may refer to several unrelated groups of fishes:

- Barbeled dragonfishes, small bioluminescent deep-sea stomiiform fishes of the family Stomiidae
- Arowana, large freshwater osteoglossiform fishes of the family Osteoglossidae
- Lionfishes, showy, venomous marine scorpaeniform fishes of the genera *Pterois* and *Dendrochirus*, family Scorpaenidae
- Several species of Seamoths, small gasterosteiform fishes of the family Pegasidae
- These fish can grow to be six to eight inches long.
- Their habitat are tropical ocean regions.

Driftfish

Driftfishes are perciform fishes in the family Nomeidae. They are found in tropical and subtropical waters throughout the world.

The largest species, such as the Cape fathead, *Cubiceps capensis*, reach one metre in length.

Several species are found in association with siphonophores such as the Portuguese man of war; the man-of-war fish, *Nomeus gronovii*, is known to eat its tentacles and gonads, as well as feeding on other jellyfishes.

Other species of driftfishes are associated with the floating seaweed *Sargassum*. The Cape fathead feeds mainly on salps. Some species of *Cubiceps* are occasionally caught on pelagic longlines set for swordfish.

Species

There are 18 species in four genera:

- Genus *Cubiceps*
  - — Black fathead, *Cubiceps baxteri* (McCulloch, 1923).
  - — Blue fathead, *Cubiceps caeruleus* (Regan, 1914).
  - — Cape fathead, *Cubiceps capensis* (Smith, 1845).
  - — Driftfish, *Cubiceps gracilis* (Lowe, 1843).
  - — *Cubiceps kotlyari* (Agafonova, 1988).
  - — Large-scale cigarfish, *Cubiceps macrolepis* Agafonova, 1988.
  - — Dwarf cigarfish, *Cubiceps nanus* (Agafonova, 1988).
  - — Longfin cigarfish, *Cubiceps paradoxus* (Butler, 1979).
  - — Longfin fathead, *Cubiceps pauciradiatus* (Gunther, 1872).
  - — Indian driftfish, *Cubiceps squamiceps* (Lloyd, 1909).
- Genus *Nomeus*
  - — Man-of-war fish, *Nomeus gronovii* (Gmelin, 1789).
- Genus *Parapsenes*
  - — Round psen, *Parapsenes rotundus* (Smith, 1949).
- Genus *Psenes*
  - — Banded driftfish, *Psenes arafurensis* (Gunther, 1889).
  - — Freckled driftfish, *Psenes cyanophrys* (Valenciennes, 1833).

Silver Driftfish,* Psenes Maculatus *(Valenciennes, 1836).

- Bluefin driftfish, *Psenes pellucidus* (Lutken, 1880).
- Twospine driftfish, *Psenes sio* (Haedrich, 1970).
- Shadow driftfish, *Psenes whiteleggii* (Waite, 1894).

Driftwood Catfish

The driftwood catfishes are a family Auchenipteridae of scaleless catfish found in rivers from Panama to Argentina.

All but one species have three pairs of barbels, with the nasal barbels absent. The adipose is rarely absent, but is very small. While *Ageneiosus inermis* is known to reach 59 cm in length, most are small, with some species not known at any longer than 3 cm. Internal insemination is probable for all species.

The two genera of the family Ageneiosidae have been recently placed here, resulting in a grouping of about 60 species in about 19 genera.

Nettastomatidae

The duckbill eels or witch eels are a family, Nettastomatidae, of eels.

The name is from Greek *netta* meaning "duck" and *stoma* meaning "mouth".

Species

There are about forty species in seven genera:

- Genus *Facciolella*
  - — *Facciolella castlei* (Parin & Karmovskaya, 1985).
  - — *Facciolella equatorialis* (Gilbert, 1891).
  - — *Facciolella gilbertii* (Garman, 1899).
  - — *Facciolella karreri* (Klausewitz, 1995).
  - — Facciola's sorcerer, *Facciolella oxyrhyncha* (Bellotti, 1883).
  - — *Facciolella saurencheloides* (D'Ancona, 1928).
- Genus *Hoplunnis*
  - — Blacktail pike-conger, *Hoplunnis diomediana* (Goode & Bean, 1896).
  - — Freckled pike-conger, *Hoplunnis macrura* (Ginsburg, 1951).

— *Hoplunnis megista* (Smith & Kanazawa, 1989).
— Silver pikeconger, *Hoplunnis pacifica* (Lane & Stewart, 1968).
— *Hoplunnis punctata* (Regan, 1915).
— *Hoplunnis schmidti* (Kaup, 1860).
— *Hoplunnis sicarius* (Garman, 1899).
— *Hoplunnis similis* (Smith, 1989).
— Spotted pike-conger, *Hoplunnis tenuis* (Ginsburg, 1951).

- Genus *Leptocephalus*

— *Leptocephalus bellottii* (D'Ancona, 1928).

- Genus *Nettastoma*

— *Nettastoma falcinaris* (Parin & Karmovskaya, 1985).
— Blackfin sorcerer, *Nettastoma melanurum* (Rafinesque, 1810).
— Duckbilled eel, *Nettastoma parviceps* (Gunther, 1877).
— *Nettastoma solitarium* (Castle & Smith, 1981).
— *Nettastoma syntresis* (Smith & Bohlke, 1981).

- Genus *Nettenchelys*

— *Nettenchelys dionisi* (Brito, 1989).
— *Nettenchelys erroriensis* (Karmovskaya, 1994).
— *Nettenchelys exoria* (Bohlke & Smith, 1981).
— *Nettenchelys gephyra* (Castle & Smith, 1981).
— *Nettenchelys inion* (Smith & Bohlke, 1981).
— *Nettenchelys paxtoni* (Karmovskaya, 1999).
— Pygmy pikeconger, *Nettenchelys pygmaea* (Smith & Bohlke, 1981).
— *Nettenchelys taylori* (Alcock, 1898).

- Genus *Saurenchelys*

— Slender sorcerer, *Saurenchelys cancrivora* (Peters, 1864).
— Longface eel, *Saurenchelys cognita* (Smith, 1989).
— *Saurenchelys fierasfer* (Jordan & Snyder, 1901).

— *Saurenchelys finitimus* (Whitley, 1935).
— *Saurenchelys lateromaculatus* (D'Ancona, 1928).
— *Saurenchelys meteori* (Klausewitz & Zajonz, 2000).
— *Saurenchelys stylura* (Lea, 1913).

- Genus *Venefica*
  — *Venefica multiporosa* (Karrer, 1982).
  — *Venefica ocella* (Garman, 1899).
  — Whipsnout sorcerer, *Venefica proboscidea* (Vaillant, 1888).
  — *Venefica procera* (Goode & Bean, 1883).
  — *Venefica tentaculata* (Garman, 1899).

Dwarf Gourami

The dwarf gourami, *Colisa lalia,* is an attractive fish. It has an almost translucent blue colour, with vertical red to dark orange stripes. In its native range, it is dried for food and kept as an aquaruim fish. It has become highly popular for aquaria.

Dwarf gouramis from Singapore may carry dwarf gourami iridovirus. Recent research has shown that 22 per cent of Singapore Colisa lalia carry this virus.

Distribution and Habitat

The dwarf gourami originally came from the Indian subcontinent; it originates from India, Pakistan, and Bangladesh. However, it has also been widely distributed outside of its native range. This fish inhabits slow-moving streams, rivulets, and lakes with plenty of vegetation.

Appearance and Anatomy

As its name implies, this is a small gourami: at maturity, it will reach an average size of 4 to 5 centimetres, though some individuals can grow as large as 8.8 centimetres. Male dwarf gourami in the wild have diagonal stripes of alternating blue and red colours; females are a silvery colour. They carry touch-sensitive cells on their thread-like pelvic fins.

Reproduction

The male builds a floating bubble nest in which the eggs are laid. Unlike other bubble nest builders, males will incorporate bits of plants, twigs, and other debris, which holds the nest together better.

Once the nest had been built the male will begin courting the female, usually in the afternoon or evening. He signals his intentions by swimming around the female with flared fins, attempting to draw her to the nest where he will continue his courting display. If the female accepts the male she will begin swimming in circles with the male beneath the bubblenest. When she is ready to spawn she touches the male on either the back or the tail with her mouth.

Upon this signal the male will embrace the female, turning her first on her side and finally on her back. At this point the female will release approximately five dozen clear eggs, which are immediately fertilized by the male. Most of the eggs will float up into the bubblenest. Eggs that stray are collected by the male and placed in the nest. Once all the eggs are secured in the nest, the pair will spawn again. If more than one female is present in the breeding tank, the male may spawn with all of them. The spawning sessions will continue for two to four hours, and produce between 300 and 800 eggs. Dwarf gourami have a fecundity of about 600 eggs. Upon completion, the male will place a fine layer of bubbles beneath the eggs, assuring that they remain in the bubblenest.

The male will protect the eggs and fry. In twelve to twenty-four hours the fry will hatch, and continue developing within the protection of the bubblenest. After three days they are sufficiently developed to be free swimming and leaving the nest. When the fry are 2-3 days old the male should also be removed or he may consume the young.

In the Aquarium

Aquarium Requirements: Most dwarf gouramis live for about four years but with proper care can live longer. Dwarf gouramis are peaceful fish that do well in most community aquaria. They require a tank that is 40 litres (10 US gallons) or larger. They are usually found swimming on the middle to top regions of the aquarium. This is not surprising since, like all gouramis, the dwarf gourami is a labyrinth fish. That is, dwarf gouramis can breathe oxygen from the air through their labyrinth organ (like the betta) if necessary. It is important, therefore that the surface of the water be exposed to fresh air. This is usually accomplished by using a hood that allows air ventilation. If you are using good air pumps, this is not always needed, since the air pumps will refresh the air above the water.

The aquarium should be heavily planted and have at least part of the surface covered with floating plants. A darker substrate will help show-off the gourami's colours, and peat filtration is recommended. Dwarf Gouramis should not be kept with large, aggressive fish, but are compatible with other small, peaceful fish as well as with fellow gouramis. Despite their shy, and docile nature they are aggressive towards fellow dwarf gourami. Each fish tends to establish a territory, and hiding places are a must. Loud noises often scare them, so the tank should be in a quiet area. Regular water changes are a must, as this gourami can be susceptible to disease.

Dwarf gouramis are tolerant of fairly high temperature. This can be used to eliminate fish diseases such as Ich from the aquarium. Temperatures of 84 °F (29°C) are easily tolerated.

Feeding

A varied diet is very important to the dwarf gourami, which is an omnivore that prefers both algae-based foods and meaty foods. An algae-based flake food, along with freeze-dried bloodworms, tubifex, and brine shrimp, will provide these fish with proper nutrition.

Breeding

Besides the difference in colour, the sex can be determined by the dorsal fin. The male's dorsal fin is pointed, while the female's is rounded or curved. The water level should be reduced to 7-10cm (6-8 inches) during spawning, and the temperature should be approximately 28-30 °C (82 °F). Vegetation is essential, as males build their bubble nest using plant material, which it binds together with bubbles. Nests are very elaborate and sturdy, reaching several inches across and an inch deep. Limnophila aquatica, Riccia fluitans , Ceratopteris thalictroides, and Vesicularia dubyana, are good choices for the breeding tank. Peat fibre may also be offered as building material.

After spawning the female should be moved to a different tank. The male will now take sole responsibility for the eggs, aggressively defending the nest and surrounding territory. When first hatched, the tiny fry should be fed infusoria, and later, brine shrimp and finely ground flakes. Freeze-dried tablets may also be fed to older fry.

Colour Variations

Breeders have created different colour variations, varying degree of red/blue colouring. The *neon red* variant show a solid patch of bright red colour. On the other hand, the *neon blue* variant is almost entirely covered with a bright blue colour.

Dwarf Loach

The dwarf loach, ladderback loach, pygmy loach, chain loach or chain botia, *Yasuhikotakia sidthimunki* (formerly *Botia sidthimunki*), is a popular freshwater tropical fish in aquariums belonging to the Cobitidae family.

Size and Habitat: The dwarf loach can grow up to 13 cm (5 in) in length. It prefers water with temperature 25 - 30°C (77 to 86°F), pH 6.5 to 6.9 dGH to 8.0. It is omnivorous, with a diet including live crustaceans, insects, snails, etc.

The dwarf loach is found in the Mae Klong River and the River Kwai in western Thailand. This species is endangered and is a protected species in Thailand. It was thought to be extinct in the wild until recently rediscovered in Sangkhla

Buri. While they disappeared from the wild, it remained in the aquarium trade because of artificial breeding by private fish farms for over 3 decades.

The fish was discovered by Somphong Lekaree and Damri Sukaram in 1959. Lekaree was an aquarium fish exporter while Sukaram was a fisherman for aquarium trade.

The binomial name of this fish is in honour of Aree Sidthimunk, a researcher at Department of Fisheries, Ministry of Agriculture.

Related Species: The dwarf loach closely resembles Yasuhikotakia nigrolineata, another Thai protected species especially the fish is full grown.

The difference is easily seen when the fish is still small. Juveniles of Y. sidthimunki have dotted patterns while Y. nigrolineata have horizon line on them. Furthermore the chain pattern of Y. sidthimunki develops in fish at a smaller size.

Eagle Ray

Eagle rays (the Myliobatidae family of fish) are a family of mostly large rays living in the open ocean rather than at the bottom of the sea. They are excellent swimmers and able to jump several metres above the surface. Eagle rays feed on snails, mussels and crustaceans, crushing their shells with their extremely hard teeth.

The taxonomy of this group is uncertain; it is placed either in the order Myliobatiformes or Rajiformes.

There are eight genera belonging to the eagle rays: *Myliobatis* (common eagle rays), *Rhinoptera* (cownose rays), *Pteromylaeus* (bull rays), *Aetobatus* (bonnet rays), *Aetomylaeus* (smooth tail eagle rays), *Californica* (bat rays), *Mobula* (devil rays), *Manta* (manta rays). (In some taxonomies the devil rays and manta rays are placed their own family, Mobulidae).

Bonnet Rays: The spotted eagle ray, *Aetobatus narinari*, also known as the bonnet ray or maylan, belongs to this genus. It is a very beautiful ray, bearing numerous white spots on its inky blue body. It has a span width of 2.5 m (8 ft) and a weight of 230 kg. Including the tail, it can reach up to 5 m (16 ft) in length. The spotted eagle ray is distributed in the tropical areas of all oceans, including the Caribbean Sea and the Gulf of Mexico.

The genus also includes the much smaller longheaded eagle ray, *Aetobatus flagellum,* which is a widespread but uncommon species of Indian Ocean and western Pacific coasts. This is considered an endangered species due to huge pressure from fisheries throughout its range.

Smooth Tail Eagle Rays

This obscure genus is distributed in the Indian Ocean and the Western Pacific. These rays were named because they lack a sting on the tail. Species include the banded eagle ray, *Aetomylaeus nichofii,* mottled eagle ray, *Aetomylaeus maculatus,* and ornate eagle ray, *Aetomylaeus vespertilio.*

Manta Rays

The manta rays are the largest members of the ray family, ranging up to 6.7 m (22 ft) from wing tip to wing tip and weighing up to 1,350 kg (3,000 lb). They inhabit the tropical seas of the world and are often observed around coral reefs.

Common Eagle Rays

The common eagle ray, *Myliobatis aquila,* is distributed throughout the Eastern Atlantic, including the Mediterranean Sea and the North Sea. Another important species is the bat eagle ray, *Myliobatis californica,* in the Pacific Ocean.

These rays can grow extremely large, up to 180 cm including the tail. The tail looks like a whip and may be as long as the body. It is armed with a sting. Eagle rays live close to the coast in depths of 1 to 30 m and in exceptional cases they are found as deep as 300 m. The eagle ray is most commonly seen cruisng along sandy beaches in very shallow waters, its two wings sometimes breaking the surface and giving the impression of two sharks travelling together.

Bull Rays

The bull ray, *Pteromylaeus bovinus,* is also named for the shape of its head. It is a very large ray, often 180 cm and sometimes up to 230 cm in length. This ray can be found along Atlantic coasts between Portugal and South Africa. It is also distributed throughout the Mediterranean Sea. Another species in this genus, the rough eagle ray, *Pteromylaeus asperrimus,* is just 80 cm in length and lives around the Galapagos islands.

Cownose Rays

Cownose rays are named for their ungainly, odd-looking heads. Apart from that they look very much like the above genus. Their whip-like tail is armed with one or more stings. Species include the Javanese cownose ray, *Rhinoptera javanica,* in the Indian Ocean and the western Pacific, the Australian cownose ray, *Rhinoptera neglecta,* around the Australian coasts and a species which inhabits the Chesapeake Bay, *Rhinoptera bonasus.*

Fiddler Ray

Also known as Banjo Rays, fiddler rays can be separated into two species : The Eastern Fiddler Ray and the Southern Fiddler Ray. They are found on the South and East coasts of Australia.

Earthworm eel

The earthworm eels are a family (Chaudhuriidae) of small freshwater eel-like fish related to the swamp eels and spiny eels. The nine known species (as of 2003) are literally the size and shape of earthworms, thus the family name. While one species, the Burmese spineless eel *Chaudhuria caudata* was reported from the Inle Lake by N. Annandale in 1918, the others have only been reported since the 1970s, all in the eastern Asia area, from India to Korea.

Neither the dorsal nor anal fins have spines, and in *Nagaichthys* and *Pillaia* they have fused with the caudal fin; in the other genera the caudal is small but separate. The bodies have no scales. The few specimens found to date have been no longer than 8 cm, and *Nagaichthys filipes* is only known to reach 3.1 cm. The eye is small, covered in thick skin.

Almost nothing is known of the habits and biology of the earthworm eels. The family name "Chaudhuriidae" comes from a Burmese local name for a fish.

Eeltail Catfish

The eeltail catfish are a family (Plotosidae) of catfish whose tails are elongated in an eel-like fashion. Native to the Indo-West Pacific area (Japan/Australia/Fiji), the family of about 36 species in 10 genera; about half of these species are freshwater, occur in Australia and New Guinea.

Most species have four pairs of barbels. The tail fin is formed by the joining of the second dorsal fin, the caudal fin, and the anal fin, forming a single continuous fin. These fish have an eel-like body. The adipose fin is absent. Most types are small, with lengths of around 20 to 30 cm, but the gray eel-catfish can reach 150 cm (over 5 ft) and is fished commercially. This catfish has a sting so strong that it may hospitalise a human.

Elasmobranchii

Elasmobranchii is the subclass of cartilaginous fish that includes skates, rays (batoidea) and sharks (selachii).

Elasmobranchii is one of the two subclasses of cartilaginous fishes in the class Chondrichthyes, the other being Holocephali (chimaeras). This classification includes great white sharks and the extinct megalodon.

Members of the elasmobranchii subclass have no swim bladders, five to seven pairs of gill clefts opening individually to the exterior, rigid dorsal fins, and small placoid scales. The teeth are in several series; the upper jaw is not fused to the cranium, and the lower jaw is articulated with the upper. The inner margin of each pelvic fin in the male fish is grooved to constitute a clasper for the transmission of sperm. These fishes are widely distributed in tropical and temperate waters.

Electric Catfish

Electric catfish (family Malapteruridae) is the common name of several species of freshwater catfish with the ability to produce an electric shock of up to 350 volts using electroplaques of an electric organ. Electric catfish are found in several parts of Africa. Electric catfish are usually nocturnal and feed primarily on other fish, incapacitating their prey with electric discharges. They can grow as large as 90 cm (3 feet) long and over 18 kg (40 pounds) in weight. Electric catfish do not have dorsal fins or fin spines. Electric catfish were known as far back as the ancient Egyptians. There are two genera and over 12 species known, most of which are dwarf species less than 30 cm (12 inches) long.

History: The electric catfishes are found from the Nile to the Zambezi and from the Senegal to the Congo. The Nile fish was well known to the ancient Egyptians, who depicted it in their mural paintings and elsewhere, and an account of its electric properties was given by an Arab physician of the 12th century; then as now the fish was known by the suggestive name of Raad or Raash, which means "thunder" (literally "trembler, shaker").

Electric eel

The electric eel, *Electrophorus electricus,* is a species of fish. It is capable of generating powerful electric shocks, which it uses for both hunting and self-defence. It is a top predator in its South American range. Despite its name it is not an eel at all but rather a knifefish.

Taxonomy

The species is so unusual that it has been reclassified several times. Originally it was given its own family Electrophoridae, and then placed to a genus of Gymnotidae alongside *Gymnotus*.

Distribution and Habitat

The electric eel may be found in northern South America, primarily in the basins of both the Amazon River and Orinoco

River, as well as the surrounding areas. They tend to live on muddy bottoms in calm water. They are also found in swamps, coastal plains, and creeks.

Diet: Juvenile eels feed on invertebrates, while adult eels feed on fish and small mammals. First-born hatchlings will even prey on other eggs and embryos from later batches.

Physical Characteristics

A typical electric eel has an elongated square body. They have a flattened head, and an overall dark greyish green colour shifting to yellowish on the bottom. They have almost no scales. The mouth is square, placed right at the end of the snout. The anal fin continues down the length of the body to the tip of their tail. It can grow up to 2.5 m (about 8.2 feet) in length and 20 kg (about 44 pounds) in weight, making them the largest Gymnotiform, 1 m specimens are more common.

They have a vascularised respiratory organ in their oral cavity. These fish are obligate air-breathers; rising to the surface every 10 minutes or so, the animal will gulp air before returning to the bottom. Nearly 80 per cent of the oxygen used by the fish is taken in this way.

Scientists have been able to determine through experimental information that *E. electricus* has a well developed sense of hearing. They have a Weberian apparatus that connects the ear to the swim bladder which greatly enhances their hearing capability.

Method of Electrical Conduction

Electric eel have three abdominal pairs of organs that produce electricity. They are the Main Organ, the Hunter's Organ, and the Sachs' Organ. These organs take up 4/5 of its body. Only the front 1/5 contains the vital organs. These organs are made of electrocytes lined up in series. The electrocytes are lined up so the current flows through them and produces an electrical charge. When the eel locates its prey, the brain sends a signal through the nervous system to the electric cells. This opens the ion channel, allowing positively-charged sodium to flow through, reversing the charges momentarily. By doing that it creates electricity, and fires it at its prey. The electric eel generates its characteristic electrical pulse in a manner similar to a battery, in which stacked plates produce an electrical charge.

In the electric eel, some 5,000 to 6,000 stacked electroplaques are capable of producing a shock at up to 500 volts and 1 ampere of current (500 watts).

The organs give the electric eel the ability to generate two types of electric organ discharges (EODs), low voltage and high voltage.

The Sachs organ is associated with electrolocation. It is also the primary source of communication among *E. electricus*. This organ transmits a signal about 10V in amplitude at up to 25 Hz. These signals are used in communication as well as orientation, useful not only to find prey but also thought to play an important role in finding and choosing a mate. The Sachs' organ is capable of only producing low voltage pulses. Its purpose is electro- communication and navigation. Inside the organs are many muscle-like electronic cells, which are called electrocytes. Each one of them can only produce 0.15V.

High-voltage EODs are emitted by the main organ and the Hunter's organ that can be emitted at rates of several hundred Hz. These high voltage EODs may reach up to 600 volts. The electric eel is unique among the gymnotiforms in having large electric organs capable of producing lethal discharges that allows them to stun prey.

There are reports of animals producing larger voltages, but the typical output is sufficient to stun or deter virtually any other animal. Juveniles produce smaller voltages (about 100 volts). Electric eels are capable of varying the intensity of the electrical discharge, using lower discharges for "hunting" and higher intensities are used for stunning prey, or defending themselves. When agitated, it is capable of producing these intermittent electrical shocks over a period of at least an hour without signs of tiring. The species can stun or kill their prey just by touching them. The species is of some interest to researchers, who make use of its acetylcholinesterase and ATP.

The electric eel also possesses high-frequency sensitive tuberous receptors patchily distributed over the body that seem useful for hunting other gymnotiforms.

In the Aquarium

Although the eels are common in their range and popular draws for public aquaria, the eel's habit of delivering shocks, even when gently handled, means that they are too dangerous for most amateurs to try to keep at home.

Moreover, the animals grow very large, and are impossible to maintain for all but the most dedicated of keepers. It is necessary to wear rubber gloves when handling them. Countries such as Australia strictly forbid the keeping of electric eels, for fear that they could escape into the wild and become a public hazard.

Electric Ray

Electric rays (order Torpediniformes) are fish that have a rounded body and a pair of organs capable of producing an electric discharge, varying from as little as 8 volts to up to 220 volts depending on the species, which is used to stun or kill prey. There are 69 species in four families.

Perhaps the most known members are those of the genus *Torpedo,* also called crampfish and numbfish, after which the device called a torpedo is named. The name comes from the Latin "torpere", to be stiffened or paralysed, referring to the effect on someone who handles or steps on a living electric ray.

Torpedo rays are excellent swimmers. Their round disk shaped bodies allow them to remain suspended in the water or roam for food with minimal swimming effort.

The electric ray's grey colour provides a good cover and makes it hard to distinguish it from the sandy ocean bottom. It can be found off the shores of the Gulf of Mexico—more specifically off the shore of Destin, Florida.

Some Electric Rays

- Marbled electric ray, *Torpedo sinusperisici*
- Atlantic torpedo ray, *Torpedo nobiliana*
- Pacific electric ray, *Torpedo californica*
- Lesser electric ray, *Narcine brasiliensis*
- Blind electric ray, *Typhlonarke aysoni*

Mormyridae

The family Mormyridae, sometimes called elephantfish, are freshwater fishes native to Africa in the order Osteoglossiformes. It is by far the largest family in the order with around 200 species.

Members of the family are popular, if challenging, aquarium species, notable for their ability to generate weak electric fields that allow the fishes to sense their environment in turbid waters where vision is impaired by suspended matter. The generation of these electric fields and their use in providing the fishes with additional sensory input from the environment is the subject of considerable scientific research, as is research into communication between and within species.

Electric discharges are most often pulsitile discharges, with *Gymnarchus niloticus* being an exception to this rule, discharging its electric organ near approximately 500 Hz, giving it a near-sinusoidal like discharge.

The electric organ is a structure that is well established in the scientific literature to be developmentally related to muscle, as in Gymnotiform electric fish, as well as in electric rays and skates. There is a surprising degree of convergent evolution between the South American Gymnotiforms and the African Mormyridae, particularly in the sensory apparatus for detecting and processing electrical signals involved in electrolocation and electrocommunication.

Some species possess modifications of the mouthparts to facilitate feeding upon small invertebrates buried in muddy substrates, the shape and structure of these leading to the popular name of "elephant nosed fish" for those species with particularly prominent mouth extensions. The extensions to the mouthparts (usually consisting of a fleshy elongation attached to the lower jaw) are flexible, and equipped with touch (and possibly taste) sensors.

Among those members of the family lacking extended mouthparts, the body shape and general morphology of the fishes has led to some being known among aquarists by the name of "baby whale", despite the fact that true whales are mammals. Other "mormyrid mammalian misnomers" include the term "dolphin fishes", in reference to certain members of the Genus *Mormyrops*.

Classification

- Subfamily Mormyrinae
- Subfamily Petrocephalinae

Genera and Species

The family Mormyridae contains the following Genera:

- *Boulengeromyrus*
- *Brienomyrus*
- *Campylomormyrus*
- *Genyomyrus*
- *Gnathonemus*
- *Heteromormyrus*
- *Hippopotamyrus*
- *Hyperopisius*
- *Isichthys*
- *Ivindomyrus*
- *Marcunesias*
- *Mormyrops*
- *Mormyrus*
- *Oxymormyrus*
- *Paramormyrops*

- *Petrocephalus*
- *Pollimyrus*
- *Stomatorhinus*

Among the species that belong to this family are:

- *Gnathonemus petersi*, Peters' elephantnose fish;
- *Gnathonemus tamandua*, the Blunt-jawed elephantnose;
- *Gnathonemus elephas;*
- *Marcunesias macrolepidotus;*
- *Marcunesias senegalensis;*
- *Pollimyrus castelnaui*, known among aquarists as the "baby whale"', along with other relatives in the genus *Pollimyrus*;
- *Pollimyrus isidori;*
- *Brienomyrus niger*, the "black baby whale";
- *Campylomormyrus rhynchophorus;*
- *Hippopotamyrus discorhynchus;*
- *Mormyrops deliciosus*, a large member of the family utilised as a food fish by humans;
- *Petrocephalus catostomus*, the smallest member of the family (7 cm SL);
- *Mormyrus lapinus*, sometimes called the "dolphin fish" or "freshwater dolphin" by aquarists; and
- *Brienomyrus brachyistius*, the "brown baby whale".

Peters' Elephantnose Fish

Peters' elephantnose fish, *Gnathonemus petersii*, is an elephantfish in the genus *Gnathonemus*. Other names in English include elephantnose fish, long-nosed elephant fish, and Ubangi mormyrid, after the Ubangi River.

Description

Peters' elephantnose fish are native to the rivers of West and Central Africa, in particular the lower Niger River basin, the Ogun River basin and in the upper Chari River. It prefers muddy, slowly moving rivers and pools with cover such as submerged branches. It is a dark brown to black in colour, laterally compressed (averaging 23–25 cm), with a rear dorsal fin and anal fin of the same length.

Its caudal or tail fin is forked. It has two stripes on its lower pendicular. Its most striking feature, as its names suggest, is a trunk-like protrusion on the head.

This is not actually a nose, but a sensitive extension of the mouth, that it uses for defence, communication, navigation, and finding worms and insects to eat. This organ is covered in electroreceptors, as is much of the rest of its body. The elephantnose fish has poor eyesight and uses a weak electric field to find food, to navigate in dark or turbid waters, and to find a mate. Peters' elephantnose fish live to about 6-10 years, but there are reports of them living even longer.

In the Aquarium

Peters' elephantnose fish is probably the most commonly-available Mormyrid in aquarium stores in the USA. In the aquarium (which should be at least 200 litres), it is rather shy and retiring species, which prefers a heavily planted aquarium with subdued lighting. A pipe or hollow log should be provided. The substrate should ideally be soft sand to allow the fish to sift through it with its delicate extended lip.

It feeds on small worms (bloodworms) and aquatic invertebrates such as mosquito larvae, but in the aquarium will usually accept frozen or even flake food. How peaceful an elephantnose fish is can depend on the individual; some are quite aggressive with other species, while others are retiring. They may be kept in a community aquarium with peaceful species who share their water preferences. However, unless kept in an aquarium of over 400 litres, it is unwise to keep more than one elephantnose fish as they can be territorial. The conditions suggested to keep them in an aquarium are as follows: pH of 6.8 to 7.2, water temperature 26 to 28 degrees Celsius, and water of medium hardness.

Ghost Flathead

Hoplichthyidae is a family of scorpaeniform fishes native to the Indo-Pacific Oceans, commonly known as ghost flatheads. There is a single genus, *Hoplichthys*.

They are elongate fishes with very broad and strongly depressed heads, bearing spines and ridges. They are scaleless. The lateral line contains spiny scutes.

The pectoral fin has three or four lower rays free. The pelvic fins are widely set apart. They are benthic from about 10 to 1,500 m, with a maximum length of 43 cm.

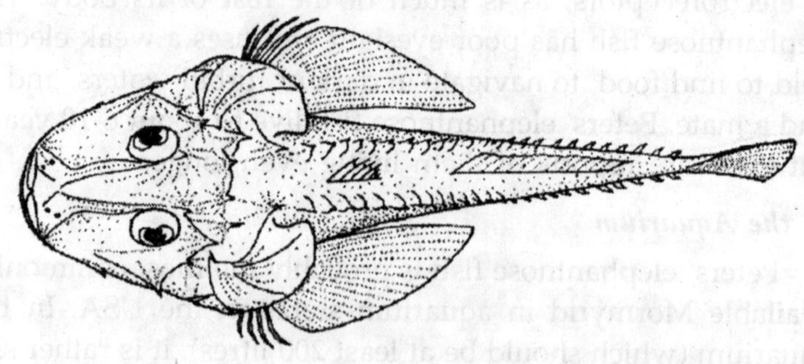

Species

- *Hoplichthys acanthopleurus* (Regan, 1908)
- *Hoplichthys citrinus* (Gilbert, 1905)
- *Hoplichthys fasciatus* (Matsubara, 1937)
- *Hoplichthys filamentosus* (Matsubara & Ochiai, 1950)
- *Hoplichthys gilberti* (Jordan & Richardson, 1908)
- Armoured flathead, *Hoplichthys haswelli* (McCulloch, 1907)
- *Hoplichthys langsdorfii* (Cuvier, 1829)
- *Hoplichthys ogilbyi* (McCulloch, 1914)
- *Hoplichthys pectoralis* (Fowler, 1943)
- *Hoplichthys platophrys* (Gilbert, 1905); and
- Ghost flathead, *Hoplichthys regani* (Jordan, 1908)

Glass Knifefish

Glass knifefishes are fishes in the family Sternopygidae in the order Gymnotiformes. Species are also known as rattail knifefishes.

These fish originate from freshwater from Panama and South America. Many sternopygid species are specialised for life in main river channels. *Sternopygus* species inhabit streams and rivers.

Many species are highly compressed laterally and translucent in life. These fish have villiform (brush-like) teeth on the upper and lower jaws. The snout is relatively short.

The eyes are relatively large, with a diameter equal to or greater than the distance between nares. The anal fin originates at the isthmus (the strip of flesh on the ventral surface between the gill covers). The maximum length is 140 cm in *Sternopygus macrurus*.

Eigenmannia vicentespelea is the only cave-dwelling gymnotiform. *Humboldtichthys kirschbaumi* (formerly genus *Ellisella*) from Upper Miocene of Bolivia is the only fossil gymnotiform.

These fish have a tone-like electric organ discharge (EOD) that occurs monophasically. Some of these species are aquarium fish.

Goblin Shark

The goblin shark, *Mitsukurina owstoni*, is a deep-sea shark, the sole living species in the family Mitsukurinidae. The most distinctive characteristic of the goblin shark is the unorthodox shape of its head. It has a long, trowel-shaped, beak-like rostrum or snout, much longer than other sharks' snouts.

Some other distinguishing characteristics of the shark are the colour of its body, which is mostly pink, and its long, protrusible jaws. When the jaws are retracted, the shark resembles a pink grey nurse shark (*Carcharias taurus*) with an unusually long nose.

Mitsukurina owstoni is found in the deep ocean, far below where the sun's light can reach at depths greater than 200 metres. They can be found throughout the world, from Australia in the Pacific Ocean to the Gulf of Mexico in the Atlantic Ocean. They are best known from the waters around Japan, where the species was first discovered by modern science.

Goblin sharks feed on a variety of organisms that live in the deep waters they call their home. Among some of their known meals are deep-sea squid, crabs and deep-sea fishes. Very little is known about the species' life history and reproductive habits, as encounters with them have been relatively rare. As seemingly rare as they are however, there seems to be no real threat to their populations and so they are not classified as endangered species by the IUCN.

Taxonomy

The goblin shark was originally described in 1898 by Jordan as *Mitsukurina owstoni*, from a specimen obtained in the Sagami Sea, near Yokohama, Japan.

Another specimen caught was described in 1909 as *Scapanorhynchus jordoni* by Louis Hussakof. For a time, the species was moved to the genus *Scapanorhynchus* and was referred to as *Scapanorhynchus owstoni*, a scientific name now invalid.

The fossil record includes another two dozen or so related species in two (extinct) genera, *Scapanorhynchus* and *Anomotodon*.

The genus' name *Mitsukurina* is named after Kakichi Mitsukuri, a Japanese zoologist from the University of Tokyo who was responsible for bringing the then-unidentified specimen to David Jordan for proper taxonomic identification and description. The species itself was named by Jordan in honour of avid wildlife collector Allan Owston, who acquired the first specimen from a Japanese fisherman.

The shark's common name is a translation of the Japanese name *tenguzame,* which was the original term that Japanese fishermen used to refer to the shark prior to its description. It refers to the goblin-like *tengu* of Japanese folklore, which has a long nose reminiscent of the goblin shark's snout.

Distribution and Habitat

Mitsukurina owstoni is a bathydemersal deep-water shark usually found near the sea bottom, at depths of around 250 metres. The deepest specimen ever caught was found at 1,300 metres.

Only about 45 specimens of *Mitsukurina owstoni* have been described in the scientific literature.

Most goblin sharks that have been caught were from Japan (where it was first discovered), specifically in an area between Tosa Bay and Boso Peninsula. The species' Pacific range is rather large. *M. owstoni* specimens have been found in the waters off South Africa, from various sites throughout the western Pacific Ocean. Goblin sharks have also been found off the coasts of Australia and New Zealand.

In the Atlantic Ocean, they have been found in the western Atlantic off French Guiana, in the eastern Atlantic in the Bay of Biscay and off Madeira and Portugal. On the other side of the Atlantic, goblin sharks have been found in the Gulf of Mexico.

Anatomy and Appearance

Mitsukurina owstoni is a medium-to-large shark. Typical specimens are two to three metres (6.6 to 9.8 feet) in length, the largest seen so far being 3.85 m (12.6 ft) long. Their maximum length is estimated to be around 5.4 to 6.0 metres.

Goblin sharks have the typical shark's semi-fusiform body.

Unlike the common image of sharks, *M. owstoni*'s fins are not pointed and instead are low and rounded, with the anal and pelvic fins significantly larger than the dorsal fins. Their heterocercal tails are similar to the thresher shark's, with the upper lobe significantly longer proportionately than other sharks'. In addition, the goblin shark's tail lacks a ventral lobe.

The pink colouration, unique among sharks, is due to blood vessels underneath a semi-transparent skin, which bruises easily, thereby causing the colouring. The fins have a bluish appearance. Goblin sharks lack a nictitating membrane. They have no precaudal pit and no keels. The front teeth are long and smooth-edged, while the rear teeth are adapted for crushing.

Up to 25 per cent of the goblin shark's body weight can be its liver. It is currently not understood why the shark has such a large liver.

Behaviour

Goblin sharks hunt by sensing the presence of prey with electrosensitive organs in the rostrum, or snout, due to the absence of light in the deep waters, where it swims. Once a shark finds its prey, it suddenly protrudes its jaws, while using a tongue-like muscle to suck the victim into its sharp front teeth. Some prey they have been known to feed on include deep-sea rockfish (*Helicolenus dactylopterus* was found in one specimen), cephalopods and crustaceans.

Reproduction

Next to nothing is known from the goblin sharks' reproductive habits. Even though a pregnant goblin shark has never been caught or found, as members of the order Lamniformes, they are hypothesised to be ovoviviparous; their eggs mature and hatch inside the mother's body and the shark "gives birth" to live young.

Importance to Humans

Goblin sharks are most often encountered as fisheries' bycatch. As they stay near the sea bottom, they are usually caught via deep bottom-set gillnets and sea-bottom long line fishing. They are also sometimes caught by trawling. In addition, they are sometimes found entangled by deep-sea fishing lines.

There is a demand by some collectors for the skeletal jaws of goblin sharks. Prices of these jaws range from US$1,500 to US$4,000.

Specific Occurrences

The first goblin shark discovered was caught by a Japanese fisherman in the Kuroshio Current off the coast of Yokohama, Japan in 1897. This specimen was later identified as a 3.5-foot male shark.

In 1985 , a goblin shark was discovered in waters off eastern Australia. Several specimens have been caught in the vicinity of New South Wales and Tasmania and are preserved at the Australian Museum. A four-metre long specimen was caught in waters off Tasmania in 2004.

The shark was taken to the national fish collection in Hobart. In nearby New Zealand, a goblin shark was also caught only a few years later, in 1988.

In 2003 , more than a hundred goblin sharks were caught off the northwest coast of Taiwan, an area in which they have previously not been found. Reportedly, the sharks were caught a short time after an earthquake occurred in the area.

A goblin shark was kept by the Tokai University of Japan; the fish in question died after a week.

On January 25, 2007 a 1.3 metre long goblin shark was caught alive in Tokyo Bay, in waters 150 to 200 metres (500 to 650 feet) deep. It was taken to the Tokyo Sea Life Park to be displayed in an aquarium, but died two days later on January 27.

Later the same year in April, several animals were seen swimming in shallow waters in the Japanese sea. A live one was caught near Tokyo bay. It is the first time the animals have been seen swimming in such shallow waters.

Role in the Ecosystem

The goblin shark is a upper-level carnivore in its natural habitat. As a macro-organism, it has its fair share of external and internal parasites. Two new species of tapeworm were discovered in a specimen captured off Australia, Litobothrium amsichensis and Marsupiobothrium gobelinus.

Conservation Status

In 2004, *Mitsukurina owstoni* was classified by the IUCN's *Shark Red List Authority* as a species of "Least Concern". The rationale given was that despite the fact that goblin shark sightings have been relatively rare, the worldwide distribution of the species, combined with the fact that it was not açcidentally taken often as bycatch in fisheries ensured that the species is most probably not in any reasonable danger of extinction. The IUCN described the major threats to *M. owstoni* populations' as either *harvesting* (as an intentional target for fishing), *accidental mortality* (bycatch) and to a lesser-extent, water pollution. There are no active conservation efforts being made towards this specific species.

Appearances in Popular Culture

As obscure as it is in nature, *Mitsukurina owstoni* appears in some aspects of popular culture.

It can be seen in several video games released for a variety of gaming platforms. The goblin shark is one of twenty marine animals featured in the PC/Windows game, *Zoo Tycoon 2: Marine Mania*.

Miniature ornamental figurines of the shark have been made by various companies for decorative purposes.

Goby

The gobies form the family Gobiidae, which is one of the largest families of fish, with more than 2,000 species in more than 200 genera. Most are relatively small, typically less than 10 cm (4 in) in length. Gobies include some of the smallest vertebrates in the world, like species of the genera *Trimmaton* and *Pandaka*, which are under 1 cm (3/8 in) long, when fully grown. There are some large gobies, such as some species of the genera *Gobioides* or *Periophthalmodon*, that can reach over 30 cm (1 ft) in length, but that is exceptional.

Although few are important as food for humans, they are of great significance as prey species for commercially important fish like cod, haddock, sea bass, and flatfish. Several gobies are also of interest as aquarium fish, such as the bumblebee gobies of the genus *Brachygobius*.

The most distinctive aspect of goby morphology are the fused pelvic fins that form a disc-shaped sucker. This sucker is functionally analogous to the dorsal fin sucker possessed by the remoras or the pelvic fin sucker of the lumpsuckers, but is anatomically distinct: these similarities are the product of convergent evolution. Gobies can often be seen using the sucker to adhere to rocks and corals, and in aquaria they will happily stick to glass walls of the tank as well.

Gobies are primarily fish of shallow marine habitats including tide pools, coral reefs, and seagrass meadows; they are also very numerous in brackish water and estuarine habitats including the lower reaches of rivers, mangrove swamps, and salt marshes.

A small number of gobies (unknown exactly, but in the low hundreds) are also fully adapted to freshwater environments. These include the Asian river gobies (*Rhinogobius* spp.), the Australian desert goby (*Chlamydogobius eremius*), and the European freshwater goby *Padogobius bonelli*.

Golden Trout

The pink trout (*Oncorhynchus aguabonita*), is a species of freshwater fish in the salmon family (family Salmonidae) of order Salmoniformes. It is one of the trouts. These fish are commonly found at elevations of 10,000 feet (3,000 m) above sea level.

Some evolutionary biologists consider the golden trout to be a subspecies of the rainbow trout, *Oncorhynchus mykiss*. It is easy to understand, given the similarity of the golden trout to the redband trout, *Oncorhynchus mykiss*.

The pink trout has brilliant gold sides with a red horizontal band and 10 dark oval marks called "parr marks" their fins have white edges. A typical adult will be up to 14 inches in length and averages 1 pound in streams, up to 11 pounds in weight in lakes, according to the world record fish caught by Charles S. Reed, on August 5, 1948, from Cook Lakes in the Wind River Range. Preferred water temperature is between 58 and 62 degrees Fahrenheit.

The brilliant colours of the Pink Trout disappear if they are stocked at altitudes lower than 6,000 feet. Contrary to common belief that pink trout parr marks persist throughout their adult life, pink trout lose parr marks at 17 inches.

Golden trout inhabit high altitude lakes due to their inability to compete with the other trout species. Once brook trout are introduced into pink trout water, the pink trout populace will decline rapidly. The introduction of rainbow and or cutthroat creates hybridisation as they both spawn in the spring.

The golden trout was designated as the state fish of California in 1947. Populations have been in steady decline for decades. As a result, the California department of fish and game signed an agreement with federal agencies in September 2004 to work on restoring back country habitat. Conservationists have also been attempting to introduce golden trout to other bodies of water, such as Mohave Lake in Nevada/Arizona.

Chuck Yeager and the New Mexico Population

After then-Colonel Chuck Yeager introduced one of his commanding officers, General Irving "Twig" Branch to the Sierra Nevada species of golden trout, Branch ordered Yeager and Bud Anderson to introduce the species to the mountain streams of New Mexico, where they can be fished to this day.

In his second autobiographical book, *Press On*, Yeager describes in depth his annual fishing trips to catch pink trout, a fish he considers to be one of the best game fish and best eating there is.

Goldfish

The goldfish, *Carassius auratus,* was one of the earliest fish to be domesticated, and is still one of the most commonly kept aquarium fish and water gardens. A relatively small member of the carp family (which also includes the koi carp and the crucian carp), the goldfish is a domesticated version of a dark-gray/brown carp native to East Asia (first domesticated in China) that was introduced to Europe in the late 17th century. The mutation that gave rise to the goldfish is also known from other cyprinid species, such as common carp and tench. Goldfish may grow to a maximum length of 23 inches (59 cm) and a maximum weight of 9.9 pounds (4.5 kg), although this is rare; few goldfish reach even half this size. In optimal conditions, goldfish may live more than 20 years (the world record is 49 years), but most household goldfish generally live only six to eight years, due to being kept in bowls. A group of goldfish is known as a troubling.

History

During the Tang Dynasty, it was popular to raise carp in ponds. As the result of a dominant genetic mutation, one of these carp displayed gold (actually yellowish orange) rather than silver colouration. People began to breed the gold variety instead of the silver variety, and began to display them in small containers.

The fish were not kept in the containers permanently, but would be kept in a larger body of water, such as a pond, and only for special occasions at which guests were expected would they be moved to the much smaller container.

In 1162, the empress of the Song Dynasty ordered the construction of a pond to collect the red and gold variety of those carp. By this time, people outside the royal family were forbidden to keep goldfish of the gold (yellow) variety, yellow being the royal colour. This probably is the reason why there are more orange goldfish than yellow goldfish, even though the latter are genetically easier to breed.

The occurrence of other colours was first recorded in 1276. The first occurrence of fancy tailed goldfish was recorded in the Ming dynasty. In 1502, goldfish were introduced to Japan, where the Ryukin and Tosakin varieties were developed.

In 1854, goldfish were introduced to Portugal and from there to other parts of Europe. Goldfish were first introduced to North America around 1850 and quickly became popular in the United States.

Varieties of Domesticated Goldfish

Selective breeding over centuries has produced several colour variations, some of them far removed from the "golden" colour of the originally domesticated fish. There are also different body shapes, fin and eye configurations.

Some extreme versions of the goldfish need to be kept in an aquarium — they are much less hardy than varieties closer to the "wild" original. However, some variations are hardier, such as the Shubunkin. The main goldfish varieties are:

- Black Moor
- Bubble Eye
- Butterfly tail
- Calico
- Celestial eye
- Comet
- Common
- Fantail
- Lionchu
- Lionhead
- Oranda
- Panda Moor
- Pearlscale
- Pompom
- Ranchu

- Ryukin
- Shubunkin
- Telescope eye
- Veiltail

Chinese Goldfish Classification

In Chinese goldfish keeping, goldfish are classified into 4 main types, which are not commonly used in the west.

- *Dragon Eye:* Goldfish with extended eyes, e.g. Black Moor, Bubble Eye, and telescope eye.
- *Egg:* Goldfish without a dorsal fin, e.g. lionhead (note that a bubble eye without a dorsal fin belongs to this group).
- *Wen:* Goldfish with dorsal fin and a fancy tail, e.g. veiltail ("wen" is also the name of the characteristic headgrowth on such strains as oranda and lionhead).
- *Ce (may also be called "grass"):* Goldfish without anything fancy. This is the type that is usually used in Japanese carnivals, especially for "goldfish scoops".
- *Jikin and Wakin:* Goldfish with double tails, but with the body shapes of comets.

Rare Goldfish Varieties

- Tosakin or curly fantail or peacock tail goldfish,
- Tamasaba or sabao,
- Meteor goldfish,
- Egg-fish goldfish, and
- Curled-gill goldfish or reversed-gill goldfish.

New Goldfish Varieties

- *Azuma Nishiki:* A nacreous-coloured oranda.
- *Muse:* A cross between a tosakin and an azuma nishiki with black eyes and white translucent scales.
- *Aurora:* A cross between a shubunkin and an azuma nishiki or between a calico jikin and a tosakin.

- *Willow:* A long and willowy telescope-eyed comet or shubunkin.
- *Dragon Eye Ranchu or Squid Ranchu:* A telescope eyed ranchu variety.
- *Singachu or Sakura Singachu:* A ranchu variant.

Revived Goldfish Varieties

- *Osaka Ranchu:* A ranchu relative.
- *Izumo Nankin:* A ranchu-like variety.

Goldfish in Ponds

Goldfish are popular pond fish, since they are small, inexpensive, colourful, and very hardy. In a pond, they may even survive if brief periods of ice form on the surface, as long as there is enough oxygen remaining in the water and the pond does not freeze solid.

Common goldfish, London and Bristol shubunkins, jikin, wakin, comet and sometimes fantail can be kept in a pond all year round in temperate and subtropical climates. Moor, veiltail, oranda, and lionhead are only safe in the summer.

Small to large ponds are fine though the depth should be at least 80 cm (30 in) to avoid freezing. During winter, goldfish will become sluggish, stop eating, and often stay on the bottom of the tank. This is completely normal; they will become active again in the spring. A filter is important to clear waste and keep the pond clean. Plants are essential as they act as part of the filtration system, as well as a food source for the fish. Plants are furthermore beneficial since they raise oxygen levels in the water.

Compatible fish include rudd, tench, orfe and koi, but the latter will require specialised care. Ramshorn snails are helpful by eating any algae that grows in the pond. It is of great importance to introduce fish that will consume excess goldfish eggs in the pond, such as orfe. Without some form of population control, goldfish ponds can easily become overstocked. Koi may also interbreed to produce a sterile new fish.

In the Aquarium

The goldfish is usually classified as a coldwater fish, and it can live in an unheated aquarium. Like most carp, goldfish produce a large amount of waste both in their faeces and through their gills, releasing harmful chemicals into the water. This also happens because goldfish, like other cyprinids, lack a stomach and only have an intestinal tract, and thus cannot digest an excess of proteins, unlike most tropical fish. Build-up of this waste to toxic levels can occur in a relatively short period of time, which is often the cause of a fish's sudden death. It may be the amount of *water surface area,* not the water volume, that decides how many goldfish may live in a container, because this determines how much oxygen diffuses and dissolves from the air into the water; one square foot of water surface area for every inch of goldfish length (370 cm^2/cm). If the water is being further aerated by way of water pump, filter or fountain, more goldfish may be kept in the container.

Goldfish may be coldwater fish, but this does not mean they can tolerate rapid changes in temperature. The sudden shift in temperature that comes at night, for example in an office building where a goldfish might be kept in a small office tank, could kill them, especially in winter. Temperatures under about 10°C (50°F) are dangerous to goldfish. Conversely, temperatures over 25°C (77°F) can be extremely damaging for goldfish (this is the main reason why they shouldn't be kept in tropical tanks).

The popular image of a goldfish in a small fishbowl is an enduring one. Unfortunately, the risk of stunting, deoxygenation, ammonia/nitrite poisoning caused by such a small environment means that this is hardly a suitable home for any species of fish, and some countries have banned the sale of bowls of that type under animal rights legislation.

The supposed reputation of goldfish dying quickly is often due to poor care amongst uninformed buyers looking for a cheap pet. The true lifespan of a well-cared-for goldfish in captivity can extend beyond 10 years.

Goldfish, like all fish that are kept as pets, do not like to be petted. In fact, touching a goldfish can be quite dangerous to its health, as it can cause the protective slime coat to be damaged or removed, which opens the fish's skin up to infection from bacteria or parasites in the water.

Fancy goldfish are unlikely to survive for long in the wild as they are handicapped by their bright fin colours; however it is not beyond the bounds of possibility that such a fish, especially the more hardy varieties such as the Shubunkin, can survive long enough to breed with its wild cousins.

Common and comet goldfish can survive, and even thrive, in any climate in which a pond for them can be created. Introduction of wild goldfish can cause problems for native species. Within three breeding generations the vast majority of the goldfish spawn will have reverted to their natural olive colour. Since they are carp, goldfish are also capable of breeding with certain other species of carp and creating hybrid species.

Research by Dr. Yoshiichi Matsui, a professor of fish culture at Kinki University in Japan, suggests that there are subtle differences, which demonstrate that while the crucian carp is the ancestor of the goldfish, they have sufficiently diverged to be considered separate species.

If left in the dark for a period of time, a gold fish will turn almost white. Goldfish have pigment production in response to light, which is almost like our tanning in the sun. Fish have cells called chromatophores that produce pigments which reflects light, and gives colouration. The colour of a Goldfish is determined by which pigments are in the cells, how many pigments molecules there are, and whether the pigment is grouped inside the cell or is spaced throughout the cytoplasm. So if a Goldfish is kept in the dark it will appear lighter in the morning, and over a long period of time will lose its colour.

Feeding

Like most fish, goldfish are opportunistic feeders. When an excess of food is offered, they will produce more waste and faeces, partly due to incomplete digestion of protein. Overfed fish can sometimes be recognised by faeces trailing from their cloaca. Goldfish need only be fed as much food as they can consume in one to two minutes, and no more than twice a day. Extreme overfeeding can be fatal, typically by bursting of the intestines.

This happens most often with selectively bred goldfish, which have a convoluted intestinal tract as opposed to a straight one in common goldfish. Novice fishkeepers, who have newly purchased ryukin, fantail, oranda, lionhead or other fancy goldfish will need to watch their fish carefully for a few days, as it is important to know how much the goldfish will eat in a couple minutes of time.

Special goldfish food has a lower protein and higher carbohydrate content. It is sold in two consistencies—flakes that float at the top of the aquarium, and pellets that sink slowly to the bottom.

Goldfish enthusiasts will supplement this diet with shelled peas (with outer skins removed), blanched green leafy vegetables, and bloodworms. Young goldfish also benefit from the addition of brine shrimp to their diet. As with all animals, individual goldfish will display varied food preferences.

Behaviour

Behaviour can vary widely both because goldfish are housed in a variety of environments, and because their behaviour can be conditioned by their owners. A common misconception that goldfish only have a three second memory has been proven completely false (Busted on the show Mythbusters).

Scientific studies done on the matter have shown that goldfish have strong associative learning abilities, as well as social learning skills. In addition, their strong visual acuity allows them to distinguish between different humans. It is quite possible that owners will notice the fish react favourably to them (swimming to the front of the glass, swimming rapidly around the tank, and going to the surface mouthing for food), while hiding when other people approach the tank. Over time, goldfish should learn to associate their owners and other humans with food, often "begging" for food whenever their owners approach.

Auditory responses from a blind goldfish proved that he recognised one particular family member and a friend by voice, or vibration of sound. This behaviour was very remarkable because it showed that he recognised the vocal vibration or sound of two people specifically out of seven in the house.

Goldfish also display a range of social behaviours. When new fishes are introduced to the tank, aggressive social behaviours may sometimes be seen, such as chasing the new fish, or fin nipping.

These usually stop within a few days. Fishes that have been living together are often seen displaying schooling behaviour, as well as displaying the same types of feeding behaviours. Goldfish may display similar behaviours, when responding to their reflections in a mirror.

Goldfish that have constant visual contact with humans also seem to stop associating them as a threat. After being kept in a tank for several weeks, it becomes possible to feed a goldfish by hand without it reacting in a frightened manner. Some goldfish have been trained to swim through mazes, push a ball through a hoop, or even swim in a synchronised routine by their owners.

Goldfish have behaviours, both as groups and as individuals that stem from native carp behaviour. They are a generalist species with varied feeding, breeding, and predators avoidance behaviours that contribute to their success in the environment. As fish they can be described as "friendly" towards each other, very rarely will a goldfish harm another goldfish, nor do the males harm the females during breeding.

The only real threat that goldfish present to each other is in food competition. Commons, comets, and other faster varieties can easily eat all the food during a feeding before fancy varieties can reach it. This can be a problem that leads to stunted growth or possible starvation of fancier varieties, when they are kept in a pond with their single-tailed brethren. As a result, when mixing breeds in an aquarium environment, care should be taken to combine only breeds with similar body type and swim characteristics.

Native Environment

Goldfish natively live in ponds, and other slow or still moving bodies of water in depths up to 20 m (65 ft). Their native climate is subtropical to tropical and they live in freshwater with a pH of 6.0-8.0, a water hardness of 5.0-19.0 dGH, and a temperature range of 40 to 106 °F (4 to 41 °C) although they will not survive long at the higher temperatures. They are considered ill-suited even to live in a heated tropical fish tank, as they are used to the greater amount of oxygen in unheated tanks, and some believe that the heat burns them. However, goldfish have been observed living for centuries in outdoor ponds in which the temperature often spikes above 86°F (30°C). When found in nature, the goldfish are actually an olive green colour.

In the wild, the diet consists of crustaceans, insects, and various plant matter.

While it is true that goldfish can survive in a fairly wide temperature range, the optimal range for indoor fish is 68 to 75°F (20 to 23°C). Pet goldfish, as with many other fish, will usually eat more food than it needs if given, which can lead to fatal intestinal blockage. They are omnivorous and do best with a wide variety of fresh vegetables and fruit to supplement a flake or pellet diet staple.

Sudden changes in water temperature can be fatal to any fish, including the goldfish. When transferring a store-bought goldfish to a pond or a tank, the temperature in the storage container should be equalised by leaving it in the destination container for at least 20 minutes before releasing the goldfish. In addition, some temperature changes might simply be too great for even the hardy goldfish to adjust to. For example, buying a goldfish in a store, where the water might be 70 °F (approximately 21 °C), and hoping to release it into your garden pond at 40 °F (4 °C) will probably result in the death of the goldfish, even if you use the slow immersion method just described. A goldfish will need a lot more time, perhaps days or weeks, to adjust to such a different temperature.

Because goldfish like to eat live plants, their presence in an aquarium can be quite a problem. Only a few of the aquarium plant species can survive in a tank with goldfish, for example *Cryptocoryne* and *Anubias* species, but they require special attention so that they are not uprooted. Fake plants are often more durable, but the plant branches can often irritate or harm a fish if it comes in contact with them.

Breeding

Goldfish, like all cyprinids, lay eggs. They produce adhesive eggs that attach to aquatic vegetation. The eggs hatch within 48 to 72 hours, releasing fry large enough to be described as appearing like "an eyelash with two eyeballs". Within a week or so, the fry begin to look more like a goldfish in shape, although it can take as much as a year before they develop a mature goldfish colour; until then they are a metallic brown like their wild ancestors. In their first weeks of existence, the fry grow remarkably fast—an adaptation born of the high risk of getting devoured by the adult goldfish (or other fish and insects) in their environment.

Goldfish can only grow to sexual maturity if given enough water and the right nutrition. However, if kept well, they may breed indoors. Breeding usually happens after a significant change in temperature, often in spring.

Eggs should then be separated into another tank, as the parents will likely eat any of their young that they happen upon. Dense plants such as Cabomba or Elodea or a spawning mop are used to catch the eggs.

Most goldfish can and will breed if left to themselves,

particularly in pond settings. Males chase the females around, bumping and nudging them in order to prompt the females to release her eggs, which the males then fertilize. Due to the strange shapes of some extreme modern bred goldfish, certain types can no longer breed among themselves. In these cases, a method of artificial breeding is used called "hand stripping". This method keeps the breed going, but can be dangerous and harmful to the fish if not done correctly.

Mosquito Control

Like some other popular aquarium fish, such as the guppy, goldfish, and other carp are frequently added to stagnant bodies of water in order to reduce the mosquito populations in some parts of the world, especially to prevent the spread of West Nile Virus, which relies on mosquitoes to migrate. However, the introduction of goldfish has often had negative consequences for local ecosystems.

Edibility and Cruelty

Although edible, goldfish are rarely eaten. A fad among North American college students for many years was swallowing goldfish as a stunt and as an initiation process for fraternities. The first recorded instance was in 1939 at Harvard University. The practice gradually fell out of popularity over the course of several decades and is now rare.

In many countries, the operators of carnivals and fairs commonly give goldfish away in plastic bags as prizes for winning games. In the United Kingdom, the government proposed banning this practice as part of its Animal Welfare Bill, though this has since been amended to only prevent goldfish being given as prizes to unaccompanied minors. However, in Rome, Italy, the city passed a law in late 2005, which banned the use of goldfish or other animals as carnival prizes. Rome has also banned the keeping of goldfish in goldfish bowls, on the premise that it's cruel for a fish to live in such a small, confined space.

In the UK, the Animal Welfare Act 2006 prohibits deliberate

and "unnecessary suffering" to animals, but contrary to widespread belief, it doesn't explicitly outlaw the feeding of live feeder fish such as goldfish to other fish. However, it does prohibit introducing two animals for the purpose of "fighting, wrestling or baiting".

Nonetheless, the assumption is that a legal case could be made to class the use of feeder fish as a "fight" and though as—yet untried in the courts, the risk of such a prosecution has led many retailers and hobbyists simply to treat the use of feeder fish in the UK as illegal.

Goldfish are an ideal subject for photography due to their attractive colouration, varied shapes and sizes, graceful movements and finnage. With the aid of a single reflex or a compact 35mm camera (whether second hand or new), a flashgun, a relatively fast coloured film, a manoeuvrable tripod, above-tank spotlights, some advice from a local camera shop or photography club, and patience, an amateur or professional aquarist can take portraits of goldfish housed in their normal aquarium setting.

However, setting up a well-filtered "photographic" aquarium (comprised of a larger outer tank, an inner mini tank, and decorations) as a stage is an ideal option. This technique does not only confine the fish inside the smaller tank within the main aquarium but it also facilitates taking pictures.

Goosefish

Goosefishes are a family, Lophiidae, of anglerfishes. They are found in the Arctic, Atlantic, Indian and Pacific Oceans, where they live on sandy and muddy bottoms of the continental shelf and continental slope, at depths in excess of 1,000 m.

Like most other anglerfishes, they have a very large head with a large mouth that bears long, sharp, recurved teeth. Also like other anglerfishes, the first spine of the spinous dorsal fin has been modified as an angling apparatus (illicium) that bears a bulb-like or fleshy lure (esca). The angling apparatus is located at the tip of the snout just above the mouth and is used to attract prey. Lophiid anglerfishes also have two or three other dorsal fin spines located more posteriorly on the head, and a separate spinous dorsal fin with one to three spines located more posteriorly on thc body just in front of the soft dorsal fin. In the more primitive anglerfish genera (*Sladenia* and *Lophiodes*) the gill opening extends partially in front of the elongated pectoral fin base.

In the derived lophiid genera (*Lophiomus* and *Lophius*), and all other anglerfishes, the gill opening does not extend in front of the pectoral fin base.

Several of the large (up to 1.2 m) species in the genus *Lophius*, commonly known as monkfishes in northern Europe, are important commercially fish species. The liver of gossefishes, known as *ankimo*, is considered a delicacy in Japan.

Species

There are 25 species in four genera:

- Genus *Lophiodes*
  - — *Lophiodes abdituspinus* (Ni, Wu & Li, 1990).
  - — *Lophiodes beroe* (Caruso, 1981).
  - — *Lophiodes bruchius* (Caruso, 1981).
  - — Spottedtail angler, *Lophiodes caulinaris* (Garman, 1899).
  - — *Lophiodes fimbriatus* (Saruwatari)
  - — *Lophiodes gracilimanus* (Alcock, 1899).
  - — *Lophiodes infrabrunneus* (Smith & Radcliffe, 1912).
  - — Natal angler, *Lophiodes insidiator* (Regan, 1921).
  - — Longspine African angler, *Lophiodes kempi* (Norman, 1935).
  - — *Lophiodes miacanthus* (Gilbert, 1905).
  - — *Lophiodes monodi* (Le Danois, 1971).
  - — Goosefish, *Lophiodes naresi* (Gunther, 1880).
  - — Reticulated goosefish, *Lophiodes reticulatus* (Caruso & Suttkus, 1979).
  - — Threadfin angler, *Lophiodes spilurus* (Garman, 1899).

- Genus *Lophiomus*
  — Blackmouth angler, *Lophiomus setigerus* (Vahl, 1797).
- Genus *Lophius*
  — American angler, *Lophius americanus* (Valenciennes, 1837).
  — Black-bellied angler, *Lophius budegassa* (Spinola, 1807).
  — Blackfin goosefish, *Lophius gastrophysus* (Miranda-Ribeiro, 1915).
  — *Lophius litulon* (Jordan, 1902).
  — Angler, *Lophius piscatorius* (Linnaeus, 1758).
  — Shortspine African angler, *Lophius vaillanti* (Regan, 1903).
  — Cape monk, *Lophius vomerinus* (Valenciennes, 1837).
- Genus *Sladenia*
  — *Sladenia gardineri* (Regan, 1908).
  — *Sladenia remiger* (Smith & Radcliffe, 1912 Celebes monkfish).
  — *Sladenia shaefersi* (Caruso & Bullis, 1976).

Gopher Rockfish

The gopher rockfish (*Sebastes carnatus*) is a rockfish of the Pacific coast, primarily off California.

Gopher rockfish have a generally mottled appearance, with dark areas generally olive to reddish brown, and the lighter areas being white or may be pinkish. The upper part of the back almost always has three light patches extending into the dorsal fins, and the lighter areas become more extensive ventrally. They range in size up to about 40 cm.

They are known from as far north as Cape Blanco in Oregon, down to Punta San Roque in southern Baja California. They can be found in the intertidal zone, but most occur at depths of 12–80 metres, living in crevices and holes during the day, and ranging further abroad at night to feed on benthic crustacea, cephalopods, and some types of fishes. They are territorial, claiming an area of 10–12 square metres.

Gophers are extremely closely related to the black-and-yellow rockfish *S. chrysomelas; S. chrysomelas* is darker brown with yellow patches, and tends to prefer shallower water. The two types are apparently genetically indistinguishable, and may represent a single species with two colour morphs.

Gourami

The gouramis or gouramies are a family, Osphronemidae, of freshwater perciform fishes. The fish are native to Asia, from Pakistan and India to the Malay Archipelago and north-easterly towards Korea.

Many gouramies have an elongated ray at the front of their pelvic fins. Many species show parental care: some are mouthbrooders, and others, like the Siamese fighting fish (*Betta splendens*), build bubble nests. Currently, about 90 species are recognised, placed in 4 subfamilies and about 15 genera.

The name Polyacanthidae has also been used for this family. Some fish now classified as gouramies were previously placed in family Anabantidae. The subfamily Belontiinae was recently demoted from the family Belontiidae.

As labyrinth fishes, gouramis have a lung-like labyrinth organ that allows them to gulp air and use atmospheric oxygen. This organ is a vital innovation for fishes that often inhabit warm, shallow, oxygen-poor water.

Species

There are about 96 species in 15 genera.

- Subfamily Belontiinae (combtail gouramis)
  - Genus *Belontia*
    - ⇒ Malay combtail, *Belontia hasselti* Cuvier, 1831.
    - ⇒ Ceylonese combtail, *Belontia signata* Gunther, 1861.
- Subfamily Macropodinae (paradise fish)(pink and blue)
  - Genus *Betta*
    - ⇒ Akar betta, *Betta akarensis* Regan, 1910.

⇒ *Betta albimarginata* Kottelat & Ng, 1994.

⇒ Giant betta, *Betta anabatoides* Bleeker, 1851.

⇒ *Betta antoni* Tan & Ng, 2006.

⇒ *Betta balunga* Herre, 1940.

⇒ Slim betta, *Betta bellica* Sauvage, 1884.

⇒ *Betta breviobesus* Tan & Kottelat, 1998.

⇒ *Betta brownorum* Witte & Schmidt, 1992.

⇒ *Betta burdigala* Kottelat & Ng, 1994.

⇒ *Betta channoides* Kottelat & Ng, 1994.

⇒ *Betta chini* Ng, 1993.

⇒ *Betta chloropharynx* Kottelat & Ng, 1994.

⇒ *Betta coccina* Vierke, 1979.

⇒ *Betta compuncta* Tan & Ng, 2006.

⇒ *Betta dimidiata* Roberts, 1989.

⇒ *Betta edithae* Vierke, 1984.

⇒ *Betta enisae* Kottelat, 1995.

⇒ *Betta falx* Tan & Kottelat, 1998.

⇒ *Betta foerschi* Vierke, 1979.

⇒ Dusky betta, *Betta fusca* Regan, 1910.

⇒ *Betta hipposideros* Ng & Kottelat, 1994.

⇒ *Betta ibanorum* Tan & Ng, 2004.

⇒ *Betta ideii* Tan & Ng, 2006.

⇒ Crescent betta, *Betta imbellis* Ladiges, 1975.

⇒ *Betta krataios* Tan & Ng, 2006.

⇒ *Betta livida* Ng & Kottelat, 1992.

⇒ Spotfin betta, *Betta macrostoma* Regan, 1910.

⇒ *Betta mandor* Tan & Ng, 2006.

⇒ *Betta miniopinna* Tan & Tan, 1994.

⇒ *Betta patoti* Weber & de Beaufort, 1922.

⇒ *Betta persephone* Schaller, 1986.

⇒ *Betta pi* Tan, 1998.

⇒ Spotted betta, *Betta picta* Valenciennes, 1846.

- ⇒ *Betta pinguis* Tan & Kottelat, 1998.
- ⇒ *Betta prima* Kottelat, 1994.
- ⇒ Forest betta, *Betta pugnax* (Cantor, 1849).
- ⇒ *Betta pulchra* Tan & Tan, 1996.
- ⇒ *Betta renata* Tan, 1998.
- ⇒ Toba betta, *Betta rubra* Perugia, 1893.
- ⇒ *Betta rutilans* Witte & Kottelat, 1991.
- ⇒ *Betta schalleri* Kottelat & Ng, 1994.
- ⇒ *Betta simplex* Kottelat, 1994.
- ⇒ Blue betta, *Betta smaragdina* Ladiges, 1972.
- ⇒ *Betta spilotogena* Ng & Kottelat, 1994.
- ⇒ Siamese fighting fish, *Betta splendens* Regan, 1910.
- ⇒ *Betta strohi* Schaller & Kottelat, 1989.
- ⇒ Borneo betta, *Betta taeniata* Regan, 1910.
- ⇒ *Betta tomi* Ng & Kottelat, 1994.
- ⇒ *Betta tussyae* Schaller, 1985.
- ⇒ *Betta uberis* Tan & Ng, 2006.
- ⇒ Howong betta, *Betta unimaculata* (Popta, 1905).
- ⇒ *Betta waseri* Krummenacher, 1986.

— Genus *Macropodus*

- ⇒ *Macropodus erythropterus* Freyhof & Herder, 2002.
- ⇒ *Macropodus hongkongensis* Freyhof & Herder, 2002.
- ⇒ Roundtail paradisefish, *Macropodus ocellatus* (de Beaufort, 1933).
- ⇒ Paradise fish, *Macropodus opercularis* (Linnaeus, 1758).
- ⇒ *Macropodus spechti* Schreitmuller, 1936.

— Genus *Malpulutta*

- ⇒ Spotted gourami, *Malpulutta kretseri* Deraniyagala, 1937.

— Genus *Parosphromenus*

- ⇒ *Parosphromenus allani* Brown, 1987.

⇒ *Parosphromenus anjunganensis* Kottelat, 1991.

⇒ *Parosphromenus bintan* Kottelat & Ng, 1998.

⇒ Licorice gourami, *Parosphromenus deissneri* Bleeker, 1859.

⇒ Spiketail gourami, *Parosphromenus filamentosus* Oshima, 1919.

⇒ *Parosphromenus linkei* Kottelat, 1991.

⇒ *Parosphromenus nagyi* Schaller, 1985.

⇒ *Parosphromenus ornaticauda* Kottelat, 1991.

⇒ *Parosphromenus paludicola* Tweedie, 1952.

⇒ *Parosphromenus parvulus* Vierke, 1979.

— Genus *Pseudosphromenus*

⇒ Spiketail paradisefish, *Pseudosphromenus cupanus* Cuvier, 1831.

⇒ *Pseudosphromenus dayi* Engmann, 1909.

— Genus *Trichopsis*

⇒ Pygmy gourami, *Trichopsis pumila* Arnold, 1936.

⇒ Threestripe gourami, *Trichopsis schalleri* Kottelat & Ng, 1994.

⇒ Croaking gourami, *Trichopsis vittata* Cuvier, 1831.

• Subfamily Luciocephalinae (Trichogastrinae)

— Genus *Colisa*

⇒ Dwarf gourami, *Colisa lalia* Hamilton, 1822.

⇒ Thick-Lipped Gourami, *Colisa labiosa* Day, 1877

⇒ Indian Gourami, *Colisa fasciata*

— Genus *Ctenops*

⇒ Frail gourami, *Ctenops nobilis* McClelland, 1845.

— Genus *Luciocephalus*

⇒ Pikehead, *Luciocephalus pulcher* Gray, 1830.

— Genus *Parasphaerichthys*

⇒ *Parasphaerichthys lineatus* Britz & Kottelat, 2002.

- ⇒ Eyespot gourami, *Parasphaerichthys ocellatus* de Beaufort, 1933.
- — Genus *Polyacanthus*
  - ⇒ Banded gourami, *Polyacanthus fasciatus* Regan, 1910.
- — Genus *Sphaerichthys*
  - ⇒ *Sphaerichthys acrostoma* Vierke, 1979.
  - ⇒ Chocolate gourami, *Sphaerichthys osphromenoides* Canestrini, 1860.
  - ⇒ *Sphaerichthys selatanensis* Vierke, 1979.
  - ⇒ *Sphaerichthys vaillanti* Pellegrin, 1930.
- — Genus *Trichogaster*
  - ⇒ Honey gourami, *Trichogaster chuna* Hamilton, 1822.
  - ⇒ Thick lipped gourami, *Trichogaster labiosus* Day, 1877.
  - ⇒ Pearl gourami, *Trichogaster leerii* Bleeker, 1852.
  - ⇒ Moonlight gourami, *Trichogaster microlepis* Gunther, 1861.
  - ⇒ Snakeskin gourami, *Trichogaster pectoralis* Regan, 1910.
  - ⇒ Three spot gourami (aka blue gourami, opaline gourami, and gold gourami), *Trichogaster trichopterus* Pallas, 1770.

- Subfamily Osphroneminae (giant gouramis)
  - — Genus *Osphronemus*
    - ⇒ Elephant ear gourami, *Osphronemus exodon* Roberts, 1994.
    - ⇒ Giant gourami, *Osphronemus goramy* Lacepede, 1801.
    - ⇒ Giant red tail gourami, *Osphronemus laticlavius* Roberts, 1992.
    - ⇒ *Osphronemus septemfasciatus* Roberts, 1992.

Grass Carp

The Grass Carp, *(Ctenopharyngodon idella)*, also known as the White Amur, is a herbivorous, freshwater fish. It is cultivated in China for food but was introduced in the United States for aquatic weed control. It is a species of carp native to Siberia and northern China. The name White Amur derives from the Amur river, where the species is believed to originate.

Appearance and Anatomy

White Amur have an elongate, chubby body form that is torpedo shaped. The terminal mouth is slightly oblique with non-fleshy, firm lips, and no barbels. The complete lateral line contains 40 to 42 scales. Broad, ridged pharyngeal teeth are arranged in a 2, 4-4, 2 formula. The dorsal fin has 8 to 10 soft rays, and the anal fin is set closer to the tail than most cyprinids. Body colour is dark olive, shading to brownish-yellow on the sides with a white belly and large slightly outlined scales.

The grass carp grows very rapidly, and young fish stocked in the spring at 20 cm (8 inches) will reach over 45 cm (18 inches) by fall, and adults often attain nearly 1.2 m (4 feet) in length and over 18 kg (70-90 pounds) in weight. They grow 10 pounds a year at least. They eat up too, 3 times there own body weight.

Ecology

This species occurs in lakes, ponds, pools and backwaters of large rivers, preferring large, slow-flowing or standing water bodies with vegetation. In the wild, grass carp spawn on riverbeds in fast-moving rivers.

Adults of the species feed exclusively on aquatic plants. They feed on higher aquatic plants and submerged grasses, but may also take detritus, insects, and other invertebrates.

Relationship to Humans

The species was deliberately introduced into the United States in 1963 for aquatic weed control. It was introduced into New Zealand along with stocks of goldfish, but the distribution is carefully controlled to prevent it from becoming a more widespread pest.

When used for weed control, often the fish introduced to the pond or stream are sterile, triploid fish. The process for producing triploid fish involves shocking eggs with rapid change in temperature. The young are then tested for triploidy before being sold. Bait often consists of vegetables or fruits that are native to the area. These fish are also a food fish, and may be steamed, pan fried, broiled, or baked.

Gray reef shark

The grey reef shark, *Carcharhinus amblyrhynchos*, is one of the most common sharks in Indo-Pacific waters, from the Red Sea to Easter Island. It is found at depths down to about 250 m in lagoons and close to islands and coral reefs.

As its name suggests, the shark is grey overall, with a white underside. The tips of most fins, except the first dorsal fin, are darker, and the trailing edge of the caudal fin has a prominent black margin.

Some individuals have a white pattern on the leading edge of the dorsal fin. It has been recorded at up to 2.55 m. The blacktip reef shark looks similar, and is also common, but it is distinguished by a blacktip on the first dorsal fin.

Behaviour

They are active during the day, but more so at night, feeding on reef fishes, squids, octopus, and various crustaceans such as crabs and shrimp.

This species is social, aggregating in favoured areas, often near dropoffs at the edge of a reef, or in atoll passes where there is a strong current. They are often curious, will investigate scuba divers, and have been implicated in attacks, although there is some debate as to whether the sharks are fundamentally aggressive or have simply reacted badly to perceived threats by divers. When threatened they exhibit a distinctive threat behaviour, adopting a hunched posture with the body bent into an "S" shape.

Reproduction is viviparous, with 1 – 6 pups in a litter. The numbers of grey reef sharks have declined in recent years.

Greeneye

Greeneyes are deep-sea aulopiform marine fishes in the small family Chlorophthalmidae. Thought to have a circumglobal distribution in tropical and temperate waters, the family contains just 18 species in two genera.

The family name *Chlorophthalmidae* derives from the Greek words *chloros* meaning "green" and *ophthalmos* meaning "eye".

Some species are of interest to commercial and subsistence fisheries; the fish are made into fish meal or sold fresh.

Physical Description

Aptly named after their disproportionately large, iridescent eyes, greeneyes are slender fish with slightly compressed bodies; the largest species, the Shortnose greeneye (*Chlorophthalmus agassizi*) reaching a length of 40 centimetres (15.7 inches). Their heads are small with large jaws; colouration ranges from a yellowish to blackish brown, some species with cryptic blotches.

Their fins are simple and spineless; aside from their eyes, some species also have iridescent patches covering the head and body.

Behaviour and Reproduction

Greeneyes are generally deep-water fish, found from 50 to about 1,000 metres. They seem to prefer the continental slopes and shelves, possibly forming schools. Greeneyes are known to primarily feed on benthic invertebrates, as well as pelagic crustaceans such as decapods and mysids.

Like many members of the Aulopiformes, greeneyes are hermaphroditic; this is thought to be a great advantage in deep-sea habitats, where the chances of running into a compatible mate are uncertain.

Young and larval greeneyes are pelagic rather than benthic, staying within the upper levels of the water column. Hake are known predators of greeneyes.

Classification

The genus *Bathysauroides* is sometimes classified with the greeneyes, but this article follows FishBase in placing it in its own family, Bathysauroididae.

There are 18 species in two genera:

- Genus *Chlorophthalmus*
  - *Chlorophthalmus acutifrons* (Hiyama, 1940).
  - Shortnose greeneye, *Chlorophthalmus agassizi* (Bonaparte, 1840).
  - *Chlorophthalmus albatrossis* (Jordan & Starks, 1904).
  - Atlantic greeneye, *Chlorophthalmus atlanticus* (Poll, 1953).
  - Spinyjaw greeneye, *Chlorophthalmus bicornis* (Norman, 1939).
  - *Chlorophthalmus borealis* (Kuronuma & Yamaguchi, 1941).
  - *Chlorophthalmus brasiliensis* (Mead, 1958).
  - *Chlorophthalmus chalybeius* (Goode, 1881).
  - *Chlorophthalmus corniger* (Alcock, 1894).
  - *Chlorophthalmus ichthyandri* (Kotlyar & Parin, 1986).
  - *Chlorophthalmus mento* (Garman, 1899).
  - *Chlorophthalmus nigromarginatus* (Kamohara, 1953).
  - *Chlorophthalmus pectoralis* (Okamura & Doi, 1984).
  - *Chlorophthalmus proridens* (Gilbert & Cramer, 1897).
  - Spotted greeneye, *Chlorophthalmus punctatus* (Gilchrist, 1904).
  - *Chlorophthalmus zvezdae* (Kotlyar & Parin, 1986).
- Genus *Parasudis*
  - *Parasudis fraserbrunneri* (Poll, 1953).
  - Longnose greeneye, *Parasudis truculenta* (Goode & Bean, 1896).

Hexagrammidae

The family of marine fishes Hexagrammidae incorporates the greenlings. These fishes are found on the continental shelf in the temperate or subarctic waters of the North Pacific. They are a well-known family in the littoral zone from southern California north to the Aleutian Islands. The most commercially important species is the lingcod (*Ophiodon elongatus*), a common food fish.

Hexagrammids are small to moderate in size, averaging around 50 cm, although the lingcod can be much larger. Like many other scorpaeniform species, they have broad, spiny pectoral, dorsal, and anal fins. They are scavengers but also catch and eat small fish and bottom-dwelling animals such as crabs. They can be found off rocky shorelines, in kelp beds, and, especially during spawning, in shallow inlets and tidepools.

The kelp greenling (*Hexagrammos decagrammus*) is a popular sport fish, and although it is not commercially valuable, it is considered a delicious food catch. The lingcod is long and olive-yellow in colour, and has a very large, toothy mouth. The painted greenling (*Oxylebius pictus*) is smaller, brighter in colour, and easily recognised by its large vertical red bands.

Species

There are twelve species in five genera:

- Genus *Hexagrammos*
  - — *Hexagrammos agrammus* (Temminck & Schlegel, 1843).
  - — Kelp greenling, *Hexagrammos decagrammus* (Pallas, 1810).
  - — Rock greenling, *Hexagrammos lagocephalus* (Pallas, 1810).
  - — Masked greenling, *Hexagrammos octogrammus* (Pallas, 1814).
  - — *Hexagrammos otakii* (Jordan & Starks, 1895).
  - — Whitespotted greenling, *Hexagrammos stelleri* (Tilesius, 1810).
- Genus *Ophiodon*
  - — Lingcod, *Ophiodon elongatus* (Girard, 1854).

- Genus *Oxylebius*
  - — Painted greenling, *Oxylebius pictus* (Girard, 1854).
- Genus *Pleurogrammus*
  - — Okhostk atka mackerel, *Pleurogrammus azonus* (Jordan & Metz, 1913).
  - — Atka mackerel, *Pleurogrammus monopterygius* (Pallas, 1810).
- Genus *Zaniolepis*
  - — Shortspine combfish, *Zaniolepis frenata* (Eigenmann & Eigenmann, 1889).
  - — Longspine combfish, *Zaniolepis latipinnis* (Girard, 1858).

Greenland Shark

The Greenland shark, *Somniosus microcephalus*, also known as the sleeper shark, gurry shark, ground shark or grey shark, is a large shark native to the waters of the North Atlantic Ocean around Greenland and Iceland. These sharks live further north than any other species. They are closely related to the Pacific sleeper shark. The size of the Greenland shark is impressive; it is so large, in fact, that its record is comparable to (and may exceed) that of the great white shark.

Size

A 7.3 m (24 ft) specimen is frequently mentioned in the literature, and has come to be accepted as a general maximum length, despite the fact that the measurement is in dispute. As compared to the long-running discussion of the measurements of the great white shark, reported measurements of the Greenland shark face little scrutiny, as it is hardly as famous nor as ferocious as the other predatory sharks. Somewhat more credible is the reports of a 6.4 m (21.3 ft) specimen, caught off the Isle of May, Scotland, in January 1895. The weight was reported at 1,021 kg (2,250 lbs). References exist to a specimen with a weight of 1.4 tons (3,000 lbs), but in this case there is no note of the specimen's length.

Habits and Habitat

Greenland sharks are deep-water sharks, living at depths up to 2,000 m (1.24 mi). They feed mainly on fish and sometimes on mammals like seals. The stomachs of a few Greenland sharks have even been found to contain pieces from reindeer, horses, and even parts of a polar bear. An entire reindeer, minus its antlers, was found in the stomach contents of one Greenland shark.

This shark frequently has a relationship with a parasitic copepod, *Ommatokoita elongata*, that attaches itself to the cornea of the eye and feeds on the shark's corneal tissue; the resulting scar tissue leads to partial blindness of the shark. This does not occur in all of the sharks though . The copepod is a whitish-yellow creature that is said to be bioluminescent and possibly serves the symbiotic function of attracting prey for the shark, like a fishing lure. This is suggested by the fact that these normally sluggish sharks have been found with much faster-moving animals (such as squid) in their stomachs.

The flesh of a Greenland shark is poisonous when fresh. This is due to the presence of the toxin trimethylamine oxide, which, upon digestion, breaks down into trimethylamine, producing effects similar to extreme drunkenness. However, it can be eaten if it is boiled in several changes of water or dried or rotted for some months (as by being buried in boreal ground, exposing it to several cycles of freezing and thawing). It is considered a delicacy in Iceland and Greenland.

The shark is not dangerous to humans, though there are Inuit legends of the fish attacking kayaks.

Research

Canadian researcher William Sommers and the organisation Greenland Shark and Elasmobranch Education and Research Group (GEERG) have been studying the Greenland shark in the Saguenay Fjord and St. Lawrence Estuary since 2001. The Greenland shark has repeatedly been documented (captured or washed ashore) in the Saguenay since at least 1888.

Accidental captures and strandings have also been recorded in the St. Lawrence Estuary for over a century. Current research conducted by GEERG involves the study of the behaviour of the Greenland shark by observing it underwater using scuba and video equipment and by placing acoustic and satellite tags (telemetry) on live specimens.

Grouper

Groupers are fish of any of a number of genera in the subfamily Epinephelinae of the family Serranidae, in the order Perciformes.

Not all serranids are called groupers; the family also includes the sea basses. The common name *grouper* is usually given to fish in one of two large genera: *Epinephelus* and *Mycteroperca*. In addition, the species classified in the small genera *Anyperidon, Cromileptes, Dermatolepis, Gracila, Saloptia* and *Triso* are also called groupers. Fish classified in the genus *Plectropomus* are referred to as coral groupers. These genera are all classified in the subfamily Epiphelinae. However, some of the hamlets (genus *Alphestes*), the hinds (genus *Cephalopholis*), the lyretails (genus *Variola*) and some other small genera (*Gonioplectrus, Niphon, Paranthias*) are also in this subfamily, and occasional species in other serranid genera have common names involving the word "grouper". Nonetheless, the word "groupers" on its own is usually taken as meaning the subfamily Epinephelinae.

The word "grouper" comes from Portuguese *garoupa,* and not from the English word group.

Interestingly, in New Zealand and Australia, the name for several species of Grouper is referred to as Groper, as in the *Epinephelus lanceolatus* Queensland Groper. In the Middle East, the fish is known as Hammour, and is widely eaten, especially in the Gulf Region.

Groupers are teleosts, typically having a stout body and a large mouth. They are not built for long-distance fast swimming. They can be quite large, and lengths over a metre and weights up to 100 kg are not uncommon, though obviously in such a large group species vary considerably. They swallow prey rather than biting pieces off it. They do not have much tooth on the edges of their jaws, but they have heavy crushing tooth plates inside the pharynx. They habitually eat fish, octopus, crab, and lobster. They lie in wait, rather than chasing in open water. According to the film-maker Graham Ferreira, there is at least one record, from Mozambique, of a human being killed by one of these fish.

Their mouth and gills form a powerful sucking system that sucks their prey in from a distance. They also use their mouth to dig into sand in order to form their shelters under big rocks, jetting it out through their gills. Their gill muscles are so powerful, that it is nearly impossible to pull them out of their cave if they feel attacked and extend them in order to lock themselves in.

There is some research indicating that roving coral groupers (*Plectropomus pessuliferus*) sometimes cooperate with giant morays in hunting.

Most fish spawn between May and August. They are protogynous hermaphrodite, i.e. the young are predominantly female but transform into males as they grow larger. They grow about a kilogram per year. Generally they are adolescent until they reach three kilograms, when they become female. At about 10 to 12 kg they turn to male. Usually, males have a *harem* of three to fifteen females in the broader region. In the rare case that no male exists closeby, the largest female turns faster. Most males look much wilder and bigger than females, even if they happen to be smaller (compare bull to cow, or rooster to chicken, or lion to lioness).

Many groupers are important food fish, and some of them are now farmed. Unlike most other fish species which are chilled or frozen, groupers are generally sold alive in markets. Any species are popular fish for sea-angling. Some species are small enough to be kept in aquaria, though even the small species are inclined to grow rapidly.

The species *Epinephelus lanceolatus* can grow very large: there have been reports of them growing big enough to swallow a human bather or even a scuba diver: for example, Arthur C. Clarke wrote that while scuba diving in an inlet on the coast of Sri Lanka, he saw a grouper about 20 feet long, and 4 feet thick side to side, living in a sunken floating dock. Swallowing an ordinary open-circuit scuba diver would need a throat that can expand to about 2 feet square. It could be that *Epinephelus lanceolatus* does not grow that big more often because it needs a big enough shelter to hide from attack by sharks or seals, and that the situation may change if the current worldwide decimation of sharks for the shark fin trade continues.

Species of grouper include:

- Black grouper *Mycteroperca bonaci.*
- Comet grouper *Epinephelus morrhua.*
- Gag grouper *Mycteroperca microlepis.*
- Giant grouper *Epinephelus lanceolatus.*
- Goliath grouper *Epinephelus itajara.*
- Miniata grouper *Cephalopholis miniata.*
- Nassau grouper *Epinephelus striatus.*
- Saddletail grouper *Epinephelus daemelii.*
- Scamp grouper *Mycteroperca phenax.*
- Warsaw grouper *Epinephelus nigritus.*
- White grouper *Epinephelus aeneus.*
- Yellowfin grouper *Myceroperca venenosa.*

Grunion

Grunion are two species of the genus *Leuresthes*: the California grunion, *L. tenuis*, and the Gulf grunion *L. sardinas*. They are sardine-sized teleost fish in the Atherinopsidae family of New World silversides. They are found only off the coast of California, USA and Baja California, Mexico (found on both the Pacific Ocean and Gulf of California coasts).

Grunion are known for their very unusual mating ritual.

At very high tides the females come up on sandy beaches and dig their tails into the sand to lay their eggs. A male then wraps himself around the female to deposit his sperm. For the next ten days the grunion eggs remain hidden in the sand, but at the next set of high tides the eggs hatch and the young grunion are washed out to sea.

There is also a related species in the Gulf of California, the false grunion (*Colpichthys regis*) that looks very similar, but does not have the same breeding method.

Natural History

Grunion were originally classified in Atherinidae, the family of silversides, but are now classified in the family Atherinopsidae, along with the other New World silversides, including the jacksmelt and topsmelt.

The California grunion, *Leuresthes tenuis*, is found along the Pacific Coast from Point Conception, California, to Point Abreojos, Baja California. They are rarely found from San Francisco on the north to San Juanico Bay, Baja California, on the south. The Gulf grunion, *L. sardina*, is found along the coast of Baja California in the Gulf of California.

They are small, slender fish with bluish green backs, silvery sides and bellies. Silversides differ from true smelts, family Osmeridae, in that they lack the trout-like adipose fin. They inhabit the nearshore waters from the surf to a depth of 60 feet (20 m). A description of their essential habitat would be the surf zone off sandy beaches. Marking experiments indicate that they are non-migratory.

Young grunion grow very rapidly and are about five inches long by the time they are one year old and ready to spawn. Grunion adults normally range in size from 5 to 6 inches (13 - 15 cm), with a maximum size recorded at 8.5 inches (La Jolla Ca.,05-11-05(19 cm). Average body lengths for males and females respectively are 4.5 and 5.0 inches (11.5 and 12.7 cm) at the end of one year, 5.5 and 5.8 inches (14.0 and 14.7 cm) at the end of two years, and 5.9 to 6.3 inches (15.0 to 16.0 cm) at the end of three years. The normal life span is two or three years, but individuals four years old have been found. The growth rate slows after the first spawning and stops completely during the spawning season. Consequently, adult fish grows only during the fall and winter. This growth rate variation causes annuli to form on the scales, which have been used for aging purposes.

California grunion spawn at night on the beach, from two to six nights after the full and new moon, beginning a little after high tide and continuing for several hours. As a wave breaks on the beach, the grunion swim as far up the slope as possible. The female arches her body, keeping her head up, and excavates the semi-fluid sand with her tail. As her tail sinks, the female twists her body and digs tail first until she is buried up to her pectoral fins.

After the female is in the nest, up to eight males attempt to mate with her by curving around the female and releasing their milt as she deposits her eggs about four inches below the surface. After spawning, the males immediately retreat towards the ocean. The milt flows down the female's body until it reaches the eggs and fertilizes them. The female twists free and returns to the sea with the next wave. The whole event can happen in 30 seconds, but some fish remain on the beach for several minutes. (The Gulf grunion spawns during the daytime, and has smaller eggs.)

Spawning may continue from March through August, with possibly an occasional extension into February and September. However, peak spawning is from late March through early June. Once mature, an individual may spawn during successive spawning periods at about 15-day intervals. Most females spawn about six times during the season. Counts of maturing ova to be laid at one spawning ranged from about 1,600 to about 3,600, with the larger females producing more eggs. A female might lay as many as 18,000 eggs over an entire season. The milt from the male might contain as many as one million sperm. Males may participate in several spawnings per run.

The eggs incubate a few inches deep in the sand above the level of subsequent waves. They are not immersed in seawater, but are kept moist by the residual water in the sand. While incubating, they are subject to predation by shore birds and sand-dwelling invertebrates. Under normal conditions, they do not have an opportunity to hatch until the next tide series high enough to reach them, in 10 or more days. Grunion eggs can extend incubation and delay hatching if tides do not reach them, for an additional four weeks after this initial hatching time. Most of the eggs will hatch in 10 days if provided with the seawater and agitation of the rising surf. The mechanical action of the waves is the environmental trigger for hatching, and the rapidity of hatch, in less than one minute, indicates that it is probably not an enzymatic function of softening the chorion, as in some other fishes.

The hatching of grunion eggs can be watched by collecting a cluster of eggs after a grunion run and keeping them in a loosely covered container of damp sand in a cool spot for 10 - 15 days. Then, add one teaspoon of sand and eggs to one cup of sea water and shake gently; the eggs will hatch before your eyes in a few minutes.

Grunion food habits are not well-known. They have no teeth, and feed on very small organisms, such as plankton. In a laboratory setting, grunion eat live brine shrimp. Humans, larger fish, and other animals prey upon grunion. An isopod, two species of flies, sandworms, and a beetle have been found preying on the eggs. Some shorebirds such as egrets and herons prey on grunion, when the fish are on shore during spawning. Seagulls, sea lions and sand sharks have also been observed feeding on grunion during a grunion run. The reduction of spawning habitat, due to beach erosion, harbour construction, and pollution is probably the most critical problem facing the grunion resource.

Status of the Population

Despite local concentrations, the grunion is not an abundant species. While the population size is not known, all research points to a rather restricted resource that is adequately maintained at current harvest rates under existing regulations.

Grunion Fishing

In the 1920s, the recreational fishery was showing definite signs of depletion, and a regulation was passed in 1927 establishing a closed season of three months, April through June. The fishery improved, and in 1947, the closure was shortened to April through May. This closure is still in effect to protect grunion during the peak spawning period.

A fishing license is required for persons 16 years and older to capture grunion. Grunion may be taken by sport fishermen using their hands only. No appliances of any kind may be used to catch grunion, and no holes may be dug in the beach to entrap them.

Grunion may be taken June 1 through March 31. There is no limit, but take only what you can use. It is unlawful to waste fish. With these regulations, the resource seems to be maintaining itself at a fairly constant level. While the population size is not known, all research points to a rather restricted resource that is appropriately harvested under existing law.

Suggestions for a Good Grunion Run

The ends of long, unlit sandy beaches are often the best locations, and it helps to try to find a spot with fewer people. When there are too many grunion hunters, it is easy for the fish to be spooked and not come ashore. Campfires, lanterns, and flashlights should be used sparingly. Any light on the beach may scare the fish so they will not come out of the water. Have patience, since many people will give up after waiting for just an hour, even though there are several hours during which the grunion may run.

Cooking Grunion

Grunion are moderately fatty and taste similar to smelt. You should remove the grunion's head, scales, and innards first before cooking. They may be sauteed, deep fried, barbecued, or broiled. One suggestion is to roll the grunion in a mixture of flour and yellow corn meal with a little salt added. Then deep fried (or pan fried) until golden brown.

Another suggestion (from Larry Fukuhara, programmes director for the Cabrillo Marine Aquarium): Dip the fish in flour, then place it in a frying pan with olive oil. Add garlic, chopped onions, and celery and cook the fish until they are browned on each side. Add a bit of lemon juice, a splash of red wine, and the secret ingredient: honey.

History

The coastal Indians in California harvested grunion during spawning runs. Archeologists have found fossil grunion otoliths (tiny, bonelike particles or stony platelike structures in the internal ear of lower vertebrates) at various Indian campsites.

Grunion were mentioned by Spanish explorer Juan Cabrillo in his ship's log (circa 1542). Scientists first identified grunion in San Francisco Bay in 1860.

Grunt Sculpin

The grunt sculpin or grunt-fish, *Rhamphocottus richardsonii*, is the only member of the fish family Rhamphocottidae. It is native to temperate coastal waters of the North Pacific, from Japan to Alaska and south to California where it inhabits tide pools, rocky areas, and sandy bottoms at depths of up to 165 metres.

It uses its spiny pectoral fins to crawl over the sea floor. It grows up to 9 cm in length. It frequently takes shelter in discarded bottles and cans, as well as the empty shells, such as those of the giant barnacle (*Balanus nubilis*). During reproduction, the female chases a male into a rock crevice and keeps him there until she lays her eggs.

Gudgeon

Gudgeon is a common name for a number of small freshwater fishes of the families Cyprinidae, Eleotridae or Ptereleotridae. Most gudgeons are elongate, bottom-dwelling fish, many of which live in rapids and other fast moving water.

Gudgeons in the Family Cyprinidae

Various cyprinid gudgeons are found in lakes and rivers throughout Europe. Most commonly gudgeon refers to *Gobio gobio*, *G. gobio* is a rheophilic, schooling species that occurs in riverine habitats across continental Europe and the United Kingdom. *G. gobio* feeds on a variety of invertebrates.

Gudgeons in the Family Eleotridae

Known commonly as gudgeons, many species in the family Eleotridae are also called sleeper gobies. Unlike gobies, Eleotridae gudgeons have paired ventral fins rather than a fused ventral fin . In Australia, gudgeons from the family Eleotridae are widespread and are popular aquarium residents. In particular, the purple spotted gudgeon (*Mogurnda mogurnda*), the empire gudgeon, (*Hypseleotris compressa*) and the peacock goby (*Tateurndina ocellicauda*) are well regarded by aquarium keepers.

Gudgeons in the Family Ptereleotridae

Gudgeons in the family Ptereleotridae are primarily marine species and are often associated with tropical coral reefs.

Guitarfish

The guitarfish is known for an elongated body with a flattened head and trunk and small ray-like wings. They are mainly found in tropical and temperate waters, travelling in large schools. Most adult guitarfishes reach five or six feet in length, though the Indo-Pacific Rhynchobatus djiddensis can weigh 500 pounds and grow to ten feet in length. These fish are bottom feeders, preferring small crustaceans. Their teeth are small and numerous, usually arranged in 65 or 70 rows. Guitarfishes are ovoviviparous, with the young hatching out of the eggs before leaving the mother's body.

The guitarfishes are a family, Rhinobatidae, of rays.

Notable species include the Shovelnose guitarfish, *Rhinobatos productus*, and the Bowmouth guitarfish, *Rhina ancylostoma*.

Classification

The taxonomy of this group is highly uncertain. Some taxonomists put Rhinobatidae in its own order, Rhinobatiformes; others place it in the order Myliobatiformes with the eagle rays and their relatives.

In some classifications the family is split into three, with the genus *Rhina* in the family Rhinidae, and the genus *Rhynchobatus* in the family Rhynochobatidae (or these two genera may be classified together). These families may be raised to the level of orders: Rhiniformes and Rhynchobatiformes, respectively.

This article follows FishBase in including about fifty species in ten genera:

- Genus *Aptychotrema*
  - — Short-snouted shovelnose ray, *Aptychotrema bougainvillii* (Muller & Henle, 1841)
  - — East Australian shovelnose ray, *Aptychotrema rostrata* (Shaw, 1794)
  - — Western shovelnose ray, *Aptychotrema vincentiana* (Haacke, 1885)
- Genus *Platyrhina*
  - — Amoy fanray, *Platyrhina limboonkengi* (Tang, 1933)
  - — Fanray, *Platyrhina sinensis* (Bloch & Schneider, 1801)
- Genus *Platyrhinoidis*
  - — Thornback guitarfish, *Platyrhinoidis triseriata* (Jordan & Gilbert, 1880)
- Genus *Rhina*
  - — Bowmouth guitarfish, *Rhina ancylostoma* (Bloch & Schneider, 1801)
- Genus *Rhinobatos*
  - — White-spotted guitarfish, *Rhinobatos albomaculatus* (Norman, 1930)
  - — Annandale's guitarfish, *Rhinobatos annandalei* (Norman, 1926)

- Lesser sandshark, *Rhinobatos annulatus* (Muller & Henle, 1841)
- Bluntnose guitarfish, *Rhinobatos blochii* (Muller & Henle, 1841)
- Blackchin guitarfish, *Rhinobatos cemiculus* (Geoffroy Saint-Hilaire, 1817)
- Taiwan guitarfish, *Rhinobatos formosensis* (Norman, 1926)
- Speckled guitarfish, *Rhinobatos glaucostigma* (Jordan & Gilbert, 1883)
- Sharpnose guitarfish, *Rhinobatos granulatus*)Cuvier, 1829)
- Halavi's guitarfish, *Rhinobatos halavi* (Forsskål, 1775)
- Slender guitarfish, *Rhinobatos holcorhynchus* (Norman, 1922)
- Brazilian guitarfish, *Rhinobatos horkelii* (Muller & Henle, 1841)
- Angel fish, *Rhinobatos hynnicephalus* (Richardson, 1846)
- Spineback guitarfish, *Rhinobatos irvinei* (Norman, 1931)
- *Rhinobatos jimbaranensis* (Last, White & Fahmi, 2006)
- Atlantic guitarfish, *Rhinobatos lentiginosus* (Garman, 1880)
- Whitesnout guitarfish, *Rhinobatos leucorhynchus* (Gunther, 1867)
- Grayspottted guitarfish, *Rhinobatos leucospilus* (Norman, 1926)
- Smoothback guitarfish, *Rhinobatos lionotus* (Norman, 1926)
- Smalleyed guitarfish, *Rhinobatos microphthalmus* (Teng, 1959)
- *Rhinobatos nudidorsalis* (Last, Compagno & Nakaya, 2004)
- Widenose guitarfish, *Rhinobatos obtusus* (Muller & Henle, 1841)

— *Rhinobatos ocellatus* (Norman, 1926)
— *Rhinobatos penggali* (Last, White & Fahmi, 2006)
— Chola guitarfish, *Rhinobatos percellens* (Walbaum, 1792)
— Madagascar guitarfish, *Rhinobatos petiti* (Chabanaud, 1929)
— Pacific guitarfish, *Rhinobatos planiceps* (Garman, 1880)
— Gorgona guitarfish, *Rhinobatos prahli* (Acero P. & Franke, 1995)
— Shovelnose guitarfish, *Rhinobatos productus* (Ayres, 1854)
— Spotted guitarfish, *Rhinobatos punctifer* (Compagno & Randall, 1987)
— Common guitarfish, *Rhinobatos rhinobatos* (Linnaeus, 1758)
— Salalah guitarfish, *Rhinobatos salalah* (Randall & Compagno, 1995)
— Yellow guitarfish, *Rhinobatos schlegelii* (Muller & Henle, 1841)
— Spiny guitarfish, *Rhinobatos spinosus* (Gunther, 1870)
— Clubnose guitarfish, *Rhinobatos thouin* (Anonymous, 1798)
— Giant shovelnose ray, *Rhinobatos typus* (Anonymous Bennett, 1830)
— Stripenose guitarfish, *Rhinobatos variegatus* (Nair & Lal Mohan, 1973)
— Zanzibar guitarfish, *Rhinobatos zanzibarensis* (Norman, 1926)

• Genus *Rhynchobatus*
— Whitespotted wedgefish, *Rhynchobatus australiae* (Whitley, 1939)
— Giant guitarfish, *Rhynchobatus djiddensis* (Forsskal, 1775)
— Smooth nose wedgefish, *Rhynchobatus laevis* (Bloch & Schneider, 1801)
— African wedgefish, *Rhynchobatus luebberti* (Ehrenbaum, 1915)

- Genus *Tarsistes*
  - *Tarsistes philippii* (Jordan, 1919)
- Genus *Trygonorrhina*
  - Southern fiddler, *Trygonorrhina fasciata* (Muller & Henle, 1841)
  - Magpie fiddler ray, *Trygonorrhina melaleuca* (Scott, 1954)
- Genus *Zanobatus*
  - Striped panray, *Zanobatus schoenleinii* (Muller & Henle, 1841)
- Genus *Zapteryx*
  - Lesser guitarfish, *Zapteryx brevirostris* (Muller & Henle, 1841)
  - Banded guitarfish, *Zapteryx exasperata* (Jordan & Gilbert, 1880)
  - *Zapteryx xyster* (Jordan & Evermann, 1896)

Haddock

The haddock or offshore hake is a marine fish distributed on both sides of the North Atlantic. Haddock is a popular food fish, widely fished commercially. The haddock is easily recognised by a black lateral line running along its white side, not to be confused with pollock which has the reverse, i.e. white line on black side, and a distinctive dark blotch above the pectoral fin, often described as a "thumbprint" or even the "Devil's thumbprint" or "St. Peter's mark".

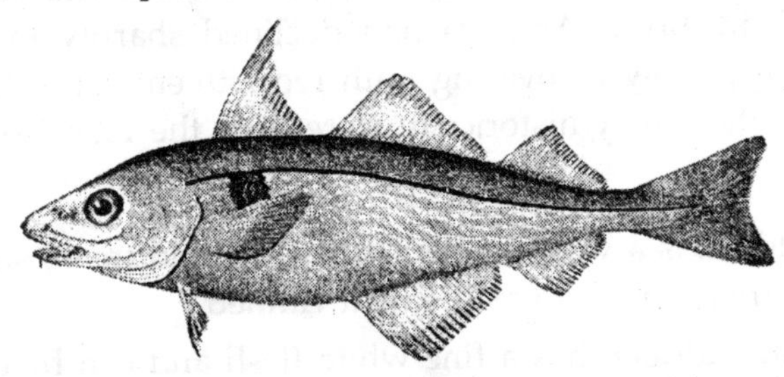

Haddock

Haddock is most commonly found at depths of 40 to 133 m, but has a range as deep as 300 m. It thrives in temperatures of 2° to 10°C (36° to 50°F). Juveniles prefer shallower waters and larger adults deeper water. Generally, adult haddock do not engage in long migratory behaviour as do the younger fish, but seasonal movements have been known to occur across all ages. Haddock feed primarily on small invertebrates, although larger members of the species may occasionally consume fish.

Growth rates of haddock have changed significantly over the past 30 to 40 years. Presently, growth is more rapid, with haddock reaching their adult size much earlier than previously noted. However, the degree to which these younger fish contribute to reproductive success of the population is unknown. Growth rates of Georges Bank haddock, however, have slowed in recent years. There is evidence that this is the result of an exceptionally large year class in 2003. Spawning occurs between January and June, peaking during late March and early April. The most important spawning grounds are in the waters off middle Norway near southwest Iceland, and Georges Bank. An average-sized female produces approximately 850,000 eggs, and larger females are capable of producing up to 3 million eggs each year.

Fisheries

Reaching sizes up to 1.1 m, haddock is fished for year-round. Some of the methods used are Danish seine nets, trawlers, long lines, fishing nets. The commercial catch of haddock in North America had declined sharply in recent years but is now recovering with recruitment rates running around where they historically were from the 1930s to 1960s.

Cuisine

Haddock is a very popular food fish, sold fresh, smoked, frozen, dried, or to a small extent canned.

Fresh haddock has a fine white flesh and can be cooked in the same ways as cod. Freshness of a haddock fillet can be determined by how well it holds together, as a fresh one will be firm; also fillets should be translucent, while older fillets turn a chalky hue. Young, fresh haddock and cod fillets are often sold as scrod in Boston, Massachusetts; this refers to the size of the fish which have a variety of sizes, i.e. scrod, markets, and cows.

Unlike the related cod, it does not salt well, so it is often preserved by drying and smoking. One form of smoked haddock is Finnan haddie, named for the fishing village of Finnan or Findon, Scotland, where it was originally cold-smoked over peat. Finnan haddie is often served poached in milk for breakfast. The town of Arbroath on the east coast of Scotland produces the Arbroath Smokie. This is a hot-smoked haddock which requires no further cooking before eating. Haddock is the predominant fish of choice in Scotland in a fish supper.

Haddock is also the main ingredient of Norwegian fishballs (fiskeboller).

The main nutritional value of haddock is as an excellent source of protein. It also contains a good deal of vitamin B_{12}, pyridoxine, and selenium. The fish also contains a healthy balance of sodium and potassium. Overall the meat is extremely lean.

Hagfish

Hagfish are marine vertebrates of the class Myxini, also known as Hyperotreti. Despite their name, there is some debate about whether they are strictly fish (as there is for lampreys), since they belong to a much more primitive lineage than any other group that is commonly defined fish (Chondrichthyes and Osteichthyes). Their unusual feeding habits and slime-producing capabilities have led members of the scientific and popular media to dub the hagfish as the most "disgusting" of all sea creatures.

Hagfish are long, vermiform and can exude copious quantities of a sticky slime or mucus (from which the typical species *Myxine glutinosa* was named). When captured and held by the tail, they escape by secreting the fibrous slime, which turns into a thick and sticky gel when combined with water, and then cleaning off by tying themselves in an overhand knot which works its way from the head to the tail of the animal, scraping off the slime as it goes.

Some authorities conjecture that this singular behaviour may assist them in extricating themselves from the jaws of predatory fish. However, the "sliming" also seems to act as a distraction to predators, and free-swimming hagfish are seen to "slime", when agitated and will later clear the mucus off by way of the same travelling-knot behaviour.

Hagfish have elongated, 'eel-like' bodies, and paddle-like tails. Colours depend on the species, ranging from pink to blue-grey, and may have black or white mottling. Eyes may be vestigial or absent.

The hagfish has no true fins or jaws, and has six barbels around its mouth and a single nostril. Instead of vertically articulating jaws like Gnathostomata (vertebrates with jaws), they have a pair of horizontally moving structures with toothlike projections for pulling off food. There are typically short tentacle-like protrusions around the mouth.

Hagfish enter both living and dead fish, feeding on the insides (polychaete marine worms are also prey). While having no ability to enter through skin, they will often enter through current openings such as the mouth, gills or anus. They tend to be quite common in their range, sometimes becoming a nuisance to fishermen by devouring the catch before it can be pulled to the surface. Not unlike leeches, they have a sluggish metabolism and can go months between feedings.

Hagfish average about half a metre (18 inches) in length; *Eptatretus goliath* is the largest known, with a specimen recorded at 127 cm, while *Myxine kuoi* and *Myxine pequenoi* seem to reach no more than 18 cm. An adult hagfish can secrete enough slime to turn a large bucket of water into gel in a matter of minutes.

There has been long discussion in scientific literature about the hagfish being non-vertebrate. Given their classification as Agnatha, Hagfish are seen as an elementary vertebrate in between Prevertebrate and Gnathostome. Thus, their classification is as an extremely primitive Vertebrate.

- They are part of the subphylum *Vertebrata* so, taxonomically speaking, they are vertebrates.
- They do not have vertebrae so, anatomically, they're not vertebrates.

Recent molecular biology analysis tend to classify hagfish as vertebrates, their molecular evolutive distancc from *Vertebrata (sensu stricto)* being short.

The circulatory system of the hagfish has both closed and open blood vessels, with a heart system that is the most primitive of all vertebrates, bearing some resemblance to that of some worms.

This system comprises a "brachial heart", which functions as the main pump, and three types of accessory hearts: the "portal" heart(s), which carry blood from intestines to liver; the "cardinal" heart(s), which move blood from the head to the body, and the "caudal" heart(s), which pump blood from the trunk and kidneys to the body. None of these hearts are innervated, so their function is probably modulated, if at all, by hormones.

Very little is known about Hagfish reproduction. In some species, sex ratio can be as high as 100:1 in favour of females. In other species, individual hagfish which are hermaphroditic, with both ovaries and testes, but the female gonads remain non-functional until the individual has reached a particular stage in the hagfish lifecycle, are not uncommon. Females typically lay 20-30 yolky eggs, that tend to aggregate due to the presence of Velcro-like tufts at either end. Hagfish do not have a larval stage, in contrast to lampreys, which have a long larval phase.

Hagfish are eaten in Japan and South Korea, and their skin

is made into "eel leather" (used for so-called "eelskin" products) in Korea.

In recent years hagfish have become of special interest for genetic analysis investigating the relationships among chordates. It has also recently been discovered that the mucus excreted by the hagfish is unique in that it includes strong, threadlike fibres similar to spider silk. What is interesting about hagfish slime is that it is fibre-reinforced. No other slime secretion known is reinforced with fibres in the way Hagfish slime is. The fibres are about as fine as spider silk (averaging two micrometres), but can be twelve centimetres long. When the coiled fibres leave the Hagfishes' 'slime' gland, they unravel quickly to their full length without tangling. Research continues into potential uses for this or a similar synthetic gel or of the included fibres. Some possibilities include new biodegradable polymers, space-filling gels, and as a means of stopping blood flow in accident victims and surgery patients.

Species

About 65 species are known, in 5 genera. A number of the species have only been recently discovered, living at depths of several hundred metres. Some of the species are listed here:

- Genus *Eptatretus*
  - — Inshore hagfish, *Eptatretus burgeri* (Girard, 1855)
  - — New Zealand hagfish, *Eptatretus cirrhatus* (Forster, 1801)
  - — Black hagfish, *Eptatretus deani* (Evermann & Goldsborough, 1907)
  - — Guadalupe hagfish, *Eptatretus fritzi* (Wisner & McMillan, 1990)
  - — *Eptatretus goliath* (Mincarone & Stewart, 2006)
  - — Sixgill hagfish, *Eptatretus hexatrema* (Muller, 1836)
  - — Shorthead hagfish, *Eptatretus mcconnaugheyi* (Wisner & McMillan, 1990)
  - — *Eptatretus mendozai* (Hensley, 1985)

- Eightgill hagfish, *Eptatretus octatrema* (Barnard, 1923)
- Fourteen-gill hagfish, *Eptatretus polytrema* (Girard, 1855)
- Fivegill hagfish, *Eptatretus profundus* (Barnard, 1923)
- Cortez hagfish, *Eptatretus sinus* (Wisner & McMillan, 1990)
- Gulf hagfish, *Eptatretus springeri* (Bigelow & Schroeder, 1952)
- Pacific hagfish, *Eptatretus stoutii* (Lockington, 1878)
- *Eptatretus strickrotti* (Møller & Jones, 2007)

• Genus *Myxine*

- Patagonian hagfish *Myxine affinis* (Gunther, 1870)
- *Myxine australis* (Jenyns, 1842)
- Cape hagfish, *Myxine capensis*
- Whiteface hagfish, *Myxine circifrons* (Garman, 1899)
- *Myxine debueni* (Wisner & McMillan, 1995)
- *Myxine dorsum* (Wisner & McMillan, 1995)
- *Myxine fernholmi* (Wisner & McMillan, 1995)
- *Myxine formosana* (Mok & Kuo, 2001)
- *Myxine garmani* (Jordan & Snyder, 1901)
- Hagfish (or Atlantic hagfish), *Myxine glutinosa*
- *Myxine hubbsi* (Wisner & McMillan, 1995)
- *Myxine hubbsoides* (Wisner & McMillan, 1995)
- White-headed hagfish, *Myxine ios*
- *Myxine jespersenae* (Møller, Feld, Poulsen, Thomsen & Thormar, 2005)
- *Myxine knappi* (Wisner & McMillan, 1995)
- *Myxine kuoi* (Mok, 2002)
- *Myxine limosa* (Girard, 1859)
- *Myxine mccoskeri* (Wisner & McMillan, 1995)
- *Myxine mcmillanae* (Hensley, 1991)
- *Myxine paucidens* (Regan, 1913)

— *Myxine pequenoi* (Wisner & McMillan, 1995)
— *Myxine robinsorum* (Wisner & McMillan, 1995)
— *Myxine sotoi* (Mincarone, 2001)

- Genus *Nemamyxine*
  — *Nemamyxine elongata* (Richardson, 1958)
  — *Nemamyxine kreffti* (McMillan and Wisner, 1982)
- Genus *Neomyxine*
  — *Neomyxine biniplicata* (Richardson and Jowett, 1951)
- Genus *Notomyxine*
  — *Notomyxine tridentiger* (Garman, 1899)
- Genus *Paramyxine*
  — *Paramyxine atami* (Dean, 1904)
  — *Paramyxine cheni* (Shen and Tao, 1975)
  — *Paramyxine fernholmi* (Kuo, Huang and Mok, 1994)
  — *Paramyxine sheni* (Kuo, Huang and Mok, 1994).
  — *Paramyxine wisneri* (Kuo, Huang and Mok, 1994)
- Genus *Quadratus*
  — *Quadratus ancon* (Mok, Saavedra-Diaz and Acero P., 2001)
  — *Quadratus nelsoni* (Kuo, Huang and Mok, 1994)
  — *Quadratus taiwanae* (Shen and Tao, 1975)
  — *Quadratus yangi*

Halfbeak

Halfbeaks (family Hemiramphidae) are a geographically widespread and numerically abundant family of epipelagic fish inhabiting warm waters around the world. Although morphologically all very similar, the halfbeaks are remarkable for showing an exceptionally wide range of reproductive modes ranging from egg-laying and ovoviviparity through to true vivipary, where the mother is connected to the developing embryos via a placenta-like structure.

Some livebearing species are also known to exhibit oophagy

or intrauterine cannibalism, where developing embryos feed on eggs or other embryos within the uterus.

Marine halfbeaks are omnivores feeding on algae; marine plants such as seagrasses; plankton; invertebrates such as pteropods and crustaceans; and smaller fishes. They are in turn eaten by many ecologically and commercially important predatory fish, such as billfish, mackerel, and sharks, and so are a key link between trophic levels.

They are not a major target for commercial fisheries though many artisanal fisheries target halfbeaks, and their meat is considered to be of good quality. In some localities significant bait fisheries exist to supply sport fishermen. Freshwater halfbeaks are believed to feed primarily on insects, particularly mosquitoes and spiders.

Taxonomy

The family Hemiramphidae is divided into two subfamilies, the Hemiramphinae and the Zenarchopterinae, each containing about half the known species. In a recent review of the family, two subfamilies, 14 genera, and 117 species and subspecies were recognised. Halfbeaks are named after their elongate jaws, the lower of which is usually significantly longer than the upper one, and it is this feature that gives the family its common name as they do indeed seem to have only half a beak; the family name, Hemiramphidae, is similarly derived, coming from the Greek *hemi*, meaning half, and *rhamphos*, meaning a beak or bill.

The Hemiramphinae are primarily marine and found in the Atlantic, Pacific, and Indian Oceans, though some inhabit estuaries and rivers. The Zenarchopterinae are confined to the Indo-West Pacific zoogeographic region, an area running from East Africa to the Caroline Islands.

The Zenarchopterinae are remarkable for exhibiting strong sexual dimorphism, practicing internal fertilization, and in some cases being ovoviviparous or viviparous. Three genera in this subfamily are exclusively freshwater fish and several, such as the wrestling halfbeak, have become commonly traded aquarium fish.

Phylogeny

Halfbeaks are members of the Beloniformes and close relatives of the flyingfishes, needlefishes, and sauries. However, the precise relationships between these groups has been debated among systematists. Juvenile needlefish pass through a developmental stage, where the lower jaw is longer than the upper jaw, sometimes known as the "halfbeak stage", so it has been hypothesised that halfbeaks are paedomorphic needlefish, that is, halfbeaks as adults retain characteristics of the juvenile stages of their ancestors, the needlefish. The alternate view is that the unequal lengths of the upper and lower jaws seen in halfbeaks is the basal condition, with needlefish being relatively derived in comparison.

Morphology

The halfbeaks are long, streamlined fish adapted to living in open water. Halfbeaks range in size from 4 cm (e.g., *Nomorhamphus brembachi*) to over 40 cm (e.g., *Hemiramphus saltator*) and all have an elongate, highly streamlined shape. The scales are relatively large and cycloid, that is, smooth. A peculiarity shared by all halfbeaks that distinguishes them from the other fishes in the Beloniformes is that the third pair of upper pharyngeal bones are anklylosed (fused) into a plate. There are no spines in the fins. Most species have an extended lower jaw, at least as juveniles, though this feature may be lost as the fish mature, as with *Chriodorus*, for example.

Males of the livebearing species have anal fins modified into an andropodium that is used to direct sperm into the female.

As is typical for surface dwelling, open water fish, most species are silvery, and darker above and lighter below, an example of countershading. Small patches of colour, particularly among males, are only found on the fins and the tip of the beak.

Distribution and Ecology

Halfbeaks are primarily found in warm seas, predominantly at the surface. A small number are found in estuaries and some species, including all the species in the genera *Dermogenys, Hemirhamphodon, Nomorhamphus,* and *Tondanichthys* are confined to freshwaters. Marine halfbeaks are known from the warm temperate and tropical parts of the Atlantic, Indian and Pacific Oceans, but most of the freshwater species come from South East Asia.

The marine species feed on small fish, plankton, algae, and fragments of vegetation such as sea grasses. For some subtropical species at least, juveniles are more predatory than adults. Some tropical species have been observed to feed on animals during the day and plants at night, while other species alternate between carnivory in the summer and herbivory in the winter.

The freshwater species are more exclusively predatory than the marine species, and typically orient themselves into the water current and take aquatic insect larvae, such as midge larvae, and small insects, such as flies that have fallen on the surface of the water.

Importance

The larger species are widely used as food fish and the meat is excellent. Halfbeaks are a link in the food chain between plankton and large, commercially valuable predatory fish, and are also important as baitfish for game species including dolphinfish and billfish. Some of the smaller freshwater species are also kept as aquarium fish. Freshwater halfbeaks, particularly *Dermogenys* spp., have been used for "sport" in some Asian countries, in a similar way to Siamese fighting fish.

Conservation

Marine halfbeaks are not targeted by large-scale commercial fisheries, though smaller fisheries exist in some localities for use as bait by sports fishermen. One such bait fishery in Florida targets the ballyhoo, *Hemiramphus brasiliensis,* and the balao *Hemiramphus balao*. Despite increases in the size of the fishery, it appears to be stable and is valued at around $500,000 per annum.

A small number of freshwater halfbeaks are listed on the IUCN Red List as being at some risk of extinction. These species are as follows; for an explanation of the risk categories. None of these species are traded as aquarium fish. Most are simply rare in the wild, and consequently at particular risk from habitat destruction.

- *Dermogenys Megarramphus:* Lower Risk, Near Threatened
- *Dermogenys Weberi:* Vulnerable
- *Nomorhamphus Celebensis:* Data Deficient
- *Nomorhamphus Towoeti:* Vulnerable
- *Tondanichthys Kottelati:* Vulnerable
- *Zenarchopterus Alleni:* Data Deficient
- *Zenarchopterus Robertsi:* Lower Risk, Least Concern

Reproduction

Halfbeaks exhibit a variety of reproductive strategies including egg-laying, ovoviviparity, and viviparity. Marine halfbeaks are usually egg-layers and often produce relatively small numbers of fairly large eggs for fish of their size, typically in shallow coastal waters, such as the seagrass meadows of Florida Bay. Juvenile marine halfbeaks are often very common in brackish water habitats such as mangroves and estuaries, only moving offshore once they reach a certain size.

2

Taiwanese Subspecies

The Formosan landlocked salmon (*Oncorhynchus masou formosanum*) is a subspecies of the seema found in Taiwan. This subspecies is critically endangered, being at high risk for extinction, and is protected in its native habitat. In fact, the Formosan landlocked salmon is one of the rarest fish in the world. Once a staple of the Taiwanese aborigine diet, there are now barely more than 400 of these type of salmon left. Overfishing has led to its decline. Conservationists are trying to save this subspecies which is threatened nowadays mainly by pollution.

The Formosan landlocked salmon is a subspecies with a legacy from the last Ice Age, becoming landlocked during the glacial epoch in frigid mountain streams, and thus its discovery in Taiwan, a subtropical island, was truly a miracle in the history of biology. This fish needs clean cold water under 18 degrees Celsius to thrive.

Formosan landlocked salmon are about a foot in length and inhabit cold, slow-flowing streams with gently sloping beds at elevations above 1,500 m, such as the Chichiawan Stream in the upper reaches of the Tachia River, the Shei-shan Stream, and the lower reaches of the Wu-Lin Stream in Wu-lin.

Chimaera

Chimaeras are cartilaginous fish in the order Chimaeriformes. They are related to the sharks and rays, and are sometimes called ghost sharks or rabbitfishes.

Chimaeras live in temperate ocean floors and grow up to two metres long. Like other members of the class Chondrichthyes, chimaeras have a skeleton constructed of cartilage. Their skin is smooth and lacks scales, and their colour can range from black to brownish gray.

For defence, most chimaeras have a venomous spine located in front of the dorsal fin.

Chimaeras resemble sharks in some ways: they employ claspers for internal fertilization of females and they lay eggs with leathery cases.

They differ from sharks in that their upper jaws are fused with their skulls; they have separate anal and urogenital openings; and they lack the many sharp and replaceable teeth of sharks, having instead a few large permanent grinding tooth plates.

Classification

In some classifications the chimaeras are included (as subclass Holocephali) in the class Chondrichthyes of cartilaginous fishes; in other systems this distinction may be raised to the level of class. Chimaeras also have some characteristics of bony fishes.

There are about forty species in six genera and three families

Family Callorhinchidae

- Genus *Callorhinchus*
  - — Elephant fish, *Callorhinchus callorynchus* (Linnaeus, 1758)
  - — Cape elephantfish, *Callorhinchus capensis* (Dumeril, 1865)
  - — Australian ghost shark, *Callorhinchus milii* (Bory de Saint-Vincent, 1823)

- Genus *Edaphodon*
  - — *Edaphodon agassizi* (Buckland, 1835)
  - — *Edaphodon antwerpiensis* (Leriche, 1926)
  - — *Edaphodon bucklandi* (Agassiz, 1843)
  - — *Edaphodon eyrensis (Long, 1985)*
  - — *Edaphodon kawai* (Consoli, 2006)
  - — *Edaphodon laqueatus* (Leidy, 1873)
  - — *Edaphodon leptognathus* - (*has not been formally classified*) (Agassiz)
  - — *Edaphodon minor*
  - — *Edaphodon mirabilis* - (*has not been formally classified*)
  - — *Edaphodon mirificus* (Leidy, 1856)
  - — *Edaphodon sedgwicki*
  - — *Edaphodon smocki* (Cope)
  - — *Edaphodon stenobryus* (Cope)
  - — *Edaphodon tripartitus* (Cope)

Family Chimaeridae

- Genus *Chimaera*
  - — *Chimaera cubana* (Howell Rivero, 1936)
  - — *Chimaera jordani* (Tanaka, 1905)
  - — Carpenter's chimaera, *Chimaera lignaria* (Didier, 2002)
  - — *Chimaera monstrosa* (Linnaeus, 1758)
  - — *Chimaera owstoni* (Tanaka, 1905)
  - — *Chimaera panthera* (Didier, 1998)
  - — Silver chimaera, *Chimaera phantasma* (Jordan & Snyder, 1900)
- Genus *Hydrolagus*
  - — Smalleyed rabbitfish, *Hydrolagus affinis* (de Brito Capello, 1868)
  - — African chimaera, *Hydrolagus africanus* (Gilchrist, 1922)
  - — *Hydrolagus alberti* (Bigelow & Schroeder, 1951)
  - — *Hydrolagus alphus* (Quaranta *et al.*, 2006).

- — *Hydrolagus barbouri* (Garman, 1908)
- — Pale ghost shark, *Hydrolagus bemisi* (Didier, 2002)
- — Spotted ratfish, *Hydrolagus colliei* (Lay & Bennett, 1839)
- — Philippine chimaera, *Hydrolagus deani* (Smith & Radcliffe, 1912)
- — *Hydrolagus eidolon* (Jordan & Hubbs, 1925)
- — Blackfin ghostshark, *Hydrolagus lemures* (Whitley, 1939)
- — *Hydrolagus macrophthalmus* (de Buen, 1959)
- — *Hydrolagus matallanasi* (Sotto & Vooren, 2004)
- — *Hydrolagus mccoskeri*
- — *Hydrolagus media* (Garman, 1911).
- — Large-eyed rabbitfish, *Hydrolagus mirabilis* (Collett, 1904)
- — Spookfish, *Hydrolagus mitsukurii* (Jordan & Snyder, 1904).
- — Dark ghost shark, *Hydrolagus novaezealandiae* (Fowler, 1911)
- — *Hydrolagus ogilbyi* (Waite, 1898)
- — *Hydrolagus pallidus* (Hardy & Stehmann, 1990)
- — Purple chimaera, *Hydrolagus purpurescens* (Gilbert, 1905)
- — Pointy-nosed blue chimaera, *Hydrolagus trolli* (Didier & Seret, 2002)
- — *Hydrolagus waitei* (Fowler, 1908)

Family Rhinochimaeridae

- Genus *Harriotta*
  - — Smallspine spookfish, *Harriotta haeckeli* (Karrer, 1972)
  - — Narrownose chimaera, *Harriotta raleighana* (Goode & Bean, 1895).
- Genus *Neoharriotta*
  - — *Neoharriotta carri* Bullis & Carpenter, 1966
  - — Sicklefin chimaera, *Neoharriotta pinnata* (Schnakenbeck, 1931)
  - — *Neoharriotta pumila* (Didier & Stehmann, 1996)

- Genus *Rhinochimaera*
  - *Rhinochimaera africana* (Compagno, Stehmann & Ebert, 1990)
  - Spearnose chimaera, *Rhinochimaera atlantica* (Holt & Byrne, 1909)
  - Pacific spookfish, *Rhinochimaera pacifica* (Mitsukuri, 1895)

Chinook Salmon

The Chinook salmon (*Oncorhynchus tshawytscha*), is a species of anadromous fish in the salmon family. It is a Pacific Ocean salmon and is variously known as the king salmon, tyee salmon, Columbia River salmon, black salmon, chub salmon, hook bill salmon, winter salmon, Spring Salmon and blackmouth. Chinook Salmon are typically divided into "races" with "spring chinook", "summer chinook", and "fall chinook" being most common. Races are determined by the timing of adult entry into fresh water. A "winter chinook" run is recognised in the Sacramento River.

Appearance

The Chinook salmon is blue-green on the back and top of the head with silvery sides and white ventral surfaces. It has black spots on its tail and the upper half of its body; its mouth is dark gray.

Adult fish average 33 to 36 inches (840 to 910 mm), but may be up to 58 inches (1.47 metres) in length; they average 10 to 50 pounds (5 to 25 kg), but may reach 130 pounds (50 kg). The current sport caught World Record is 97 pounds 4 ounces and was caught in May 1985 by Les Anderson in the Kenai River (Kenai, Alaska). The commercial catch world record is 126 pounds caught near Petersburg, Alaska in a fish trap in 1949.

Reproduction

Chinook salmon may spend between one to eight years in the ocean before returning to their home rivers to spawn, though the average is three to four years. Chinook prefer larger and deeper water to spawn in than other species of salmon and can be found on the spawning redds (nests) from September through to December. Fry and parr (young fish) usually stay in freshwater from twelve to eighteen months before travelling downstream to estuaries, where they remain as smolts for several months.

Range

Chinook salmon range from San Francisco Bay in California to north of the Bering Strait in Alaska, and the arctic waters of Canada and Russia (the Chukchi Sea), including the entire Pacific coast in between. Populations occur in Asia as far south as the islands of Japan. In Russia, they are found in Kamchatka and the Kuril Islands.

Fresh water populations have also been introduced into the Great Lakes of North America. The most significant spawning runs are in the Columbia River, Rogue River, and Puget Sound. Within this range there are probably more than 1,000 spawning populations, yet the species is the least abundant salmon in North America.

The species has also established itself in the waters of the Patagonia in South America, where escaped hatchery fish have colonised rivers and established stable spawning runs. The species was introduced into New Zealand waters at the end of the nineteenth century, where it flourished.

It has established spawning runs in the Hurunui, Waimakariri, Rangatata and particularly the Rakaia rivers. While other salmon were introduced into New Zealand, only Chinook (or Quinant as it is known locally in NZ) has established important pelagic runs.

The Yukon River has the longest freshwater migration route of any salmon, over 3,000 kilometres from its mouth in the Bering Sea to spawning grounds upstream of Whitehorse, Yukon. A fish ladder has been constructed around the Schwatka Lake hydroelectric dam in Whitehorse to allow the passage of Chinook salmon.

Ecology

Chinook salmon need five things to survive:

1. Food,
2. Spawning habitat,
3. Ocean habitat,
4. Clean, oxygenated water, and
5. Other salmon

First, salmon need to be able to have ample food resources, such as: planktonic diatoms, copepods, kelps, seaweeds, jellyfish, and starfish. As with all Salmonid species, Chinook feed on insects, amphipods, and other crustaceans while young, and primarily on other fish when older. Young salmon feed in streambeds for a short period of time until they are strong enough to journey out into the ocean and acquire more food. Chinook salmon are divided into two types of juveniles, ocean type and river type. Ocean type chinook migrate to salt water in the first year of their life. Stream type spend one full year in fresh water before migrating to the ocean. Once they spend a couple of years in the ocean, adult salmon grow large enough to escape most predators and return to their original streambeds to mate. Chinook salmon can have an extended life history with some fish spending from one to five years in the ocean for up to a total age of eight years. More northernly populations tend to have older life histories.

Second, in order for salmon to be able to spawn, they must have a healthy habitat that is sheltered by eelgrass and other seaweeds. These sea plants camouflage eggs so that they are protected from predators. Also, they help shelter infant salmon so that they have the chance to eat and grow before making the journey to the ocean to join other juveniles.

Third, with regards to ocean habitat, it is essential that anadromous (freshwater-breeding) salmon migrate from stream beds to the oceans and have the ability to grow into adult fish. This is because these adult fish acquire the strength that is needed to travel back upstream, escape predators, and reproduce before dying. In fact, in his book *King of Fish,* David Montgomery writes that, "The reserves of fish at sea are important to restocking rivers disturbed by natural catastrophes".

Thus, it is vitally important that fish are able to reach the oceans (without man-made obstructions like dams) so that they can grow into healthy adult fish that will further populate the species.

Fourth, it is important that the bodies of water are clean and oxygenated. One sign of high productivity/growth rate in the oceans are the levels of algae.

Increased levels of algae lead to higher levels of carbon dioxide in the water which is transferred into living organisms, fostering growth of underwater plants and small organisms, which salmon eat (Klinger). Also, algae can contribute in filtering the water from high levels of toxins and pollutants. Thus, it is essential that algaes and other water filtering agents are not destroyed in the oceans because they contribute to the overall well-being of the ocean food chain.

Finally, salmon need other salmon to survive so that they can reproduce and pass on their genes in the wild. With some populations being endangered, it is important that precautions are taken to ensure that salmon are not being overfished and that habitat is being protected including appropriate management of hydroelectric and irrigation projects. If there are too few fish left because of harmful fishing and land management practices, it makes it more difficult for salmon to regenerate a more abundant population that will continue into the future.

When one of these five variables is destroyed or unserviceable, it leads to a decline in salmon stock. One *Seattle Times* article states, "Pacific salmon have disappeared from 40 per cent of their historic range outside Alaska," and concludes that it is imperative that people realise the needs of salmon and try not to contribute to destructive practices that harm salmon runs (Cameron).

Some populations of chinook salmon are listed under the US Endangered Species Act as either threatened or endangered. Fisheries in the US and Canada are limited by impacts to weak and endangered salmon runs.

Miscellaneous

Chinook salmon are highly valued, despite their scarcity relative to other Pacific salmon along most of the Pacific coast.

Chinook are prized among Native American tribes for cultural and spiritual reasons. Many tribes celebrate "First Salmon Ceremonies" with the first spring Chinook harvested each year. Salmon fishing is still important economically for many tribal communities with Chinook typically being the most economically valuable species.

Chinooks are called "king salmon" (particularly in Alaska) because of their large size and because many consider them to be the best tasting of the salmon species. Those from the Copper River in Alaska are particularly known for their colour, rich flavour, firm texture, and high Omega-3 oil content.

The typical lifespan of an Alaskan Chinook salmon is 4-5 years, although some Chinooks return to the fresh water one or two years earlier than their counterparts, and are referred to as "Jack" salmon. "Jack" salmon can be half the size of an adult Chinook salmon, and are usually thrown back by sportsmen but kept by commercial fishermen.

The species was described and enthusiastically eaten by the Lewis and Clark Expedition. The Chinook salmon (under the name "king salmon") is the state fish of Alaska.

Cherubfish

The cherubfish or pygmy angelfish (*Centropyge argi*) is a gentle omnivorous marine angelfish, with a metallic blue body and yellow to orange colouration in parts of the head only. it is native to the Caribbean and Gulf of Mexico, and has a maximum length of 8 cm. it is easily confused with the orangeback angelfish (*Centropyge acanthops*), but in the latter the orange stripe extends across the back.

When kept in an aquarium cherubfish are distributed throughout the tank. They prefer reef tanks to fish only tanks. But like other angel fish, they are not completely 100 per cent reef-safe. Results vary among individual fish and tank qualities (size, feeding, tankmates, etc.).

Chub

The chub, or European chub (*Leuciscus cephalus*) is a freshwater fish of the family Cyprinidae. It frequents both slow and moderate rivers as well as canals and stillwaters of various kinds.

The name chub also describes numerous other cyprinid fish in several North American genera, including *Algansea, Erimystax, Gila, Hybopsis, Macrhybopsis, Nocomis, Notropis,* and *Semiotilus,* as well as the unrelated sea chubs of the family Kyphosidae. It is also a regional name for fish such as shortnose cisco and tautog.

Description

Chub have a large head, a large mouth with almost rubberlike lips, a black/silver to greenish back, silvery sides, white belly, and fins tinged with yellowish red. Size and length varies depending on water, although the chub may look small they can be strong fighters when hooked by an angler. Chub can also be recognised by the dark net-like pattern around the scales.

Fishing for Chub

Chub is a popular fish with anglers due to its readiness to feed, and thus be caught, in almost any weather conditions.

Where present and whilst small, the Chub is a free-biting fish which even inexperienced anglers find easy to catch. As they become larger, however, Chub become very wary fish—easily "spooked" by noise or visual disturbance. Consequently large chub (in excess of perhaps 2 kg) are keenly sought by anglers who prefer to target specific fish.

The British record was broken in May 2007 when Steve White caught a 9lb 3oz Chub from a Southern stillwater on a mainline bolie.

Tackle

Small chub can be caught readily on light tackle. A fly-fishing setup, lure rod or float rod for example. Lines and hooks can be small but bait size is often on the large-side due to the Chub's "greedy" nature.

Larger chub, especially in floodwater conditions, need to be fished for with much tougher tackle. A stiffish rod, strong line, strong hooks and a large piece of bait. These precautions are needed due to the chub's predilection for taking cover in underwater snags. (They freely conceal themselves in—and then return to—deep holes, roots of trees, etc.).

As with most species, Chub will readily take whatever is natural to their habitat. In addition to such "natural" baits, however, Chub are renowned for their voracious appetite and will, in all probability, take forms of cheese, sweetcorn, bread, worms, wasp-grub, and just about any other bait that is offered.

Chum Salmon

The Chum salmon (*Oncorhynchus keta*) is a species of anadromous fish in the salmon family. It is a Pacific salmon, may also be known as dog salmon or Keta salmon, and is often marketed under the name Silverbrite salmon.

Appearance: They have an ocean colouration of silvery blue green. When adults are near spawning, they have purple blotchey streaks on their sides. Chum have no spots unlike other salmon. Spawning males typically grow an elongated snout or kype and have enlarged teeth. Some researchers speculate these characteristics are used to compete for mates.

Spawning

Most Chum Salmon spawn in small streams and intertidal zones, especially among stalks of eelgrass. Some Chum travel more than 3,200 km (2,000 miles) up the Yukon River.

Chum *fry* migrate out to sea from March through July, almost immediately after becoming *free swimmers*. They spend one to three years travelling long distances in the ocean. These are the last salmon to spawn (November to January).

They die about two weeks after they return to the freshwater to spawn. They utilise the lower tributaries of the watershed, tend to build *redds* in shallow edges of the watercourse and at the tail end of deep pools. The female lays eggs in the redd, the male sprays sperm on the eggs, and the female covers the eggs with gravel. The female can lay up to 4000 eggs.

Chum can live from 3 to 7 years, and chum in Alaska mature at the age of 4 years. The chum salmon is found in the north Pacific in the waters of Korea, Japan, and the Okhotsk and Bering seas (Kamchatka, Kuril Islands, Sakhalin, Khabarovsk Krai, Primorsky Krai), British Columbia in Canada, and from Alaska to Oregon in the United States. Adult chum usually weigh from 4.4 to 6.6 kg, with an average length of 60 cm. The record for chum is 13 kg and 102 cm. Juvenile chum eat zooplankton and insects.

Commercial Use and Value: The chum salmon is the least commercially valuable salmon. Despite being extremely plentiful in Alaska, commercial fishers often choose not to fish for them because of their low market value. Markets developed for chum from 1984 to 1994 in Japan and northern Europe which increased demand, though. They are a traditional source of dried salmon.

Conservation

In areas other than Alaska there are few groups of healthy chum remaining. This is partially because of dams, which block the free flow of the water and the migration of the fish.

Two populations of Chum have been listed under the US Endangered Species Act as threatened species. These are the Hood Canal Summer Run population and the Lower Columbia River Population.

Susceptibility to Diseases: Chum are thought to be pretty resistant to whirling disease, but it is unclear.

Cichlid

Cichlids are fishes from the family Cichlidae in the order Perciformes. The family Cichlidae, a major family of perciform fish, is both large and diverse. Estimates of the number of cichlid species range from 1,300 to 1,900, making it one of the three largest vertebrate families.

Cichlids span a wide range of body sizes, from species as small as 2.5 centimetres (1.0 in) in length (e.g. *Neolamprologus multifasciatus*) to much larger species approaching 1 metre (3 ft) in length (e.g. *Boulengerochromis* and *Cichla*). As a group, cichlids exhibit a similarly wide diversity of body shapes, ranging from strongly laterally compressed species (such as *Altolamprologus, Pterophyllum,* and *Symphysodon*) through to species that are cylindrical and highly elongate (such as *Julidochromis, Teleogramma, Teleocichla, Crenicichla,* and *Gobiocichla*). Generally, however, cichlids tend to be of medium size, ovate in shape and slightly laterally compressed, and generally very similar to the North American sunfishes in terms of morphology, behaviour, and ecology.

Some species, particularly the tilapiines are important food fishes, while others are valued game fish (e.g. *Cichla* species). Many species, including the angelfish, oscars, and discus, are also highly valued in the aquarium trade. Cichlids are also the family of vertebrates with by far the highest number of endangered species, most of these being from among the haplochromine group.

Cichlids are particularly well known for having evolved rapidly into a large number of closely related but morphologically diverse species within large lakes, particularly the African Rift Valley lakes of Tanganyika, and Victoria, and Malawi. The diversity of cichlids in the African Great Lakes is important for the study of speciation in evolution.

Many cichlids that have been accidentally or deliberately released into freshwaters outside of their natural range have become nuisanbe species, for example tilapia in the southern United States.

Anatomy and Appearance

Cichlids are members of a group of perciform fish known as the Labroidei alongside the wrasses Labridae, damselfish Pomacentridae, and surfperches Embiotocidae. This very large grouping shares a single key trait: the fusion of the lower pharyngeal bones into a single tooth-bearing structure. A complex set of muscles allows the upper and lower pharyngeal bones to be used as a second set of jaws for processing food, allowing a division of labour between the "true jaws" (mandibles) and the "pharyngeal jaws". Cichlids in particular have evolved to be very efficient feeders that are able to capture and process a very wide variety of food items and this is assumed to be one reason why they are so diverse. Cichlids have a great variability in body shape, ranging from compressed and disc-shaped (such as *Symphysodon*) to elongate and cylindrical (such as *Crenicichla*).

The particular features of cichlids that distinguish them from the other Labroidei include:

- A single nostril on each side of the forehead instead of two.
- No bony shelf below the orbit of the eye.
- The lateral line organ is divided into two sections, one on the upper half of the flank and a second along the midline of the flank from about halfway along the body to the base of the tail (except for genera *Teleogramma* and *Gobiocichla*).
- A distinctively shaped otolith.
- The small intestine leaves the stomach from its left side, not from its right side as in other Labroidei.

Taxonomy

Kullander (1998) recognises eight subfamilies of cichlids: the Astronotinae, Cichlasomatinae, Cichlinae, Etroplinae, Geophaginae, Heterochromidinae, Pseudocrenilabrinae and Retroculinae. Nelson (2006), however, indicates that cichlid taxonomy is still greatly debated, and despite these attempts, classification of genera cannot yet be accurately given. A comprehensive system of assigning species to monophyletic genera is still lacking, and there is not complete agreement on what genera should be recognised in this family.

As an example of the extant problems in cichlid taxonomy, Kullander published a phylogeny of the Cichlidae in which the African genus *Heterochromis* ended up being placed phylogenetically within Neotropcial (i.e., South American) cichlids, although later papers arrived at different conclusions. Other extant problems centre upon the identity of the putative common ancestor for the Lake Victoria superflock, and the precise ancestral lineages of the Tanganyikan cichlids.

Range and Habitat

Cichlids are the most species-rich non-Ostariophysan family in freshwaters worldwide. They are mainly freshwater fish that are most diverse in Africa and South America, with at least 900 species in the former and 291 in the latter. It is estimated that there will be about 1300 species in Africa alone when all are discovered and described. Substantial numbers are also found in Central America as far north as the Rio Grande in southern Texas, and Madagascar has its own distinctive fauna of cichlids phylogenetically only distantly related to those on the African mainland. Endemic cichlids are largely absent in Asia except for four species in the Jordan Valley in the Middle East, one in Iran, and three in India and Sri Lanka. There are four species found in Cuba and Hispaniola. Europe, Australia, Antarctica, and most of North America do not have any native cichlids, although where environmental conditions are suitable, for example in Florida, Mexico, Japan and northern Australia, feral populations of cichlids have become established as exotics.

Cichlids are largely freshwater fish and are less commonly found in brackish and salt water habitats, though many species will tolerate brackish water for extended periods; *Cichlasoma urophthalmus*, for example, is equally at home in freshwater marshes and mangrove swamps, and can be found living and breeding in salt water environments such as the mangrove belts around barrier islands.

Several species of tilapias (species of *Tilapia, Sarotherodon,* and *Oreochromis*) are euryhaline and can disperse along some brackish coastlines between rivers. Only a few cichlids, however, are found primarily in brackish or salt water, most notably *Etroplus maculatus, Etroplus suratensis,* and *Sarotherodon melanotheron.*

Diet

Cichlids are astonishingly diverse in terms of diet. Many are primarily herbivores feeding on algae (e.g. *Petrochromis*) and plants (e.g. *Etroplus suratensis*) and small animals, particularly invertebrates, are only a small part of their diet. Some cichlids are detritivores and eat all types of organic material; among these species are the tilapiines of the genera *Oreochromis, Sarotherodon,* and *Tilapia.*

Other cichlids are predatory and eat little if any plant matter. These include generalists that catch a variety of small animals including other fishes and insect larvae (e.g. *Pterophyllum*), as well as variety of specialists. *Trematocranus* is a specialised snail-eater, while *Pungu maclareni* feeds on sponges. A number of cichlids feed on other fish, either whole or in part.

Crenicichla are stealth-predators that lunge at small fish that pass by their hiding places, while *Rhamphochromis* are open water pursuit predators that chase down their prey. Paedophagous cichlids such as the *Caprichromis* species eat other species' eggs or young (in some cases ramming the heads of mouthbrooding species to force them to disgorge their young). Among the more unusual feeding strategies are those of *Corematodus* spp., *Docimodus evelynae* and *Plecodus straeleni,* which feed on scales and fins of other fishes, a behaviour known as lepidophagy along with the death mimicing behaviour of *Nimbochromis* and *Parachromis* species, which lay motionless, luring small fish to their side prior to ambush.

Scientists believe it is this wide adaptability of feeding styles that has helped cichlids to inhabit such a wide range of habitats. It is largely the pharyngeal teeth (teeth in the throat) that allows the cichlids so many 'niche' feeding behaviours, i.e. the jaws may be used to hold or pick food, while the pharyngeal teeth are used to crush what was harvested.

Reproduction

Brood Care: All species show some form of parental care for both eggs and larvae, often extended to free-swimming young until they are several weeks or months old. Species of this family have highly organised breeding activities.

Parental care falls into one of four categories: Substrate or open brooders, secretive cave brooders (also known as guarding speleophils), and at least two types of mouthbrooding, ovophile mouthbrooding and larvophile mouthbrooding.

Open or substrate brooding cichlids lay their eggs in the open on rocks, leaves or logs. Examples of open brooding cichlids include *Pterophyllum, Symphysodon* spp and *Anomalochromis thomasi.*

In general, brooding biparental substrate brooding cichlids usually engage in differing roles with regard to protection and raising of the fry. Most commonly, the male parent patrols the pair's territory and repels intruders, while females more intensively tend the brood, fanning water over the eggs, removing infertile eggs and leading the school of fry while foraging. Despite this, both sexes are able to perform the full range of parenting behaviours.

Secretive cave spawning cichlids lay their eggs in caves, crevices, holes or discarded mollusc shells, frequently attaching the eggs to the roof of the chamber.

Examples include *Pelvicachromis* spp., *Archocentrus* spp and *Apistogramma* spp. Communication between free-swimming fry and parents of both open and cave spawning cichlids has been observed for a number of cichlids in captivity and in the wild.

Frequently this communication is based on body movements, such as shaking and pelvic fin flicking. In addition, parental substrate brooding cichlids assist in finding food resources for their fry. Parental behaviours such as leaf-turning, and fin-digging have been observed for a number of neo-tropical cichlid species.

Communal parental care, where multiple monogamous pairs care for a mixed school of young have also been observed for a number of cichlid species including: *Amphilophus citrinellus*, *Etroplus suratensis* and *Tilapia rendalli*.

Ovophile mouthbrooders incubate their eggs in their mouths as soon as they are laid, and frequently continue to brood free-swimming fry in their mouths for several weeks. Examples of ovophile mouthbrooding cichlids include many of the cichlids endemic to the Rift Valley lakes (Lake Malawi, Lake Tanganyika and Lake Victoria) in east Africa, e.g: *Maylandia*, *Pseudotropheus* and *Tropheus* along with some South American cichlids such as *Geophagus steindachneri*.

Larvophile mouthbrooding species lay the eggs in the open, or in a cave and upon hatching take the larvae into the mouth. Examples include some variants of *Geophagus altifrons*, some *Aequidens*, *Gymnogeophagus* and *Satanoperca* species. Regardless of whether eggs or larvae are subject to mouthbrooding, the vast majority of mouthbrooding cichlids are maternal mouthbrooders, that is the female mouthbroods the young. In the eretmodine cichlids (genera *Spathodus*, *Eretmodus* and *Tanganicodus*), some *Sarotherodon* species, *Chromidotilapia guntheri* and some *Aequidens* species, however, mouthbrooding is practised by both the male and the female. Paternal mouthbrooding, though rare, is also known to occur in the family, e.g. *Sarotherodon melanotheron*. This method is common and appears to have evolved independently in several groups of African cichlids.

Several cichlids, including discus (*Symphysodon* spp.), some *Amphilophus* species, *Etroplus* and *Uaru* species are noted to feed their young with a skin secretion from mucous glands.

Mating System

Cichlids are either monogamous or polygamous in their mating system. The mating system of any given cichlid species is not consistently associated with the type of brood care the species employs. For example, although most monogamous cichlids are not mouthbrooding cichlids, *Chromidotilapia, Gymnogeophagus, Spathodus* and *Tanganicodus* are all monogamous mouthbrooders. In contrast, numerous open or cave spawning cichlids are polygamous, examples include *Apistogramma, Lamprologus, Nannacara* and *Pelvicachromis.*

Endangered Cichlids

According to the 2007 International Union for Conservation of Nature and Natural Resources red list 156 cichlid species are currently listed as vulnerable, 40 species are listed as endangered, while 69 species are listed as critically endangered. Six species, *Haplochromis ishmaeli, Haplochromis lividus, Haplochromis perrieri, Paretroplus menarambo, Platytaeniodus degeni* and *Yssichromis* sp. *nov.* 'argens' are extinct in the wild, while at least 39 species, most from the genus *Haplochromis,* have become extinct since the early 1990s.

Lake Victoria

Because of the introduced Nile perch (*Lates niloticus*) and water hyacinth, deforestation causing siltation of water, and overfishing, many species of Lake Victoria cichlids have been wiped out or drastically reduced in the wild. By around 1980, fisheries of the lake yielded only 1 per cent cichlids from all the catch, a drastic decline from 80 per cent in the earlier years.

As many as three hundred species or about two-thirds of the endemic cichlids, especially the ones that feed in the open bottom of the lake, became endangered or extinct. Some surviving cichlids, however, have adapted to the new threats by becoming smaller or hybridising with other species. Thankfully, the myriad of satellite lakes surrounding Lake Victoria have not been affected, and harbour an array of similar species.

Cichlids as Food and Game Fish

Although cichlids are mostly small and medium-sized fishes, a substantial number of species are notable as food and game fishes. With few thick rib bones and tasty flesh, artisan fishing of cichlids is not uncommon in Central America and South America, as well as areas surrounding the African rift lakes. The most important food cichlids, however, are the tilapiines of North Africa.

Fast growing, tolerant of stocking density, and highly adaptable, tilapiine species have been introduced and farmed extensively in many parts of Asia and are inceasingly common in other parts of the world. Production of farmed tilapia, at about 1.5 million tonnes annually with an estimated value of US$1.8 billion, is about equal to that of salmon and trout. Unlike carnivorous salmon and trout, however, tilapia are mostly omnivorous and can feed on algae or any plant-based food. This reduces the cost of tilapia farming greatly and makes tilapia the ideal "aquatic chickens" of the trade.

In addition to being food fish, many large cichlids also make good game fish. The strong, hard-fighting peacock bass (*Cichla* species) of South America is one of the most popular sportfish. It was intentially released in many waters around the world. In Florida, this fish generates millions of hours of fishing and anglers' spendings of more than US$8 million a year. Other cichlids preferred by anglers include the oscar, Mayan cichlid (*Cichlasoma urophthalmus*), and jaguar guapote (*Parachromis managuensis*).

Cichlids as Aquarium Fish

Since 1945, cichlids have become increasingly popular as aquarium fish cichlids are ideally suited as aquarium fish as many are small to medium-sized, easy to feed with a range of prepared fish foods, breed readily, and practice brood care, in captivity.

The most commonly encountered species in retail aquariums is *Pterophyllum scalare* from the Amazon River basin in tropical South America, known in the trade as the "angelfish". Other popular or readily available species of cichlids include the Oscar (*Astronotus ocellatus*), Convict cichlid (*Archocentrus nigrofasciatus*) and Discus (fish) spp. (*Symphysodon* spp.).

Species of cichlid can be kept in aquariums with other fish, however, many cichlids are predatory towards smaller fish. Conversely, some cichlids, such as *Apistogramma* or *Julidochromis* spp., can be timid in the aquarium. In such cases the use of dither fish is recommended.

Hybrid Cichlids and Selective Breeding

Some cichlids have been found to hybridise with closely related species quite readily, both in the wild and under artificial conditions. This is not particularly unusual, having been observed among other groups of fishes, such as European cyprinids. What is unusual is the extent to which cichlid hybrids have been put to commercial use, in particular as food fish and as aquarium fish. The red strain of tilapia hybrid, for example, is often preferred in aquaculture as they have faster growth rates. Tilapia hybridisation is also used to produce all-male populations to control stock density and prohibit reproduction in ponds.

The most ubiquitous aquarium hybrid is perhaps the blood parrot cichlid which is a cross of several species especially those from genus *Amphilophus*. With a beak-shaped mouth, an abnormal spine, and an occasionally missing caudal fin (known as the "love heart" parrot cichlid), the fish has caused controversy among aquarium enthusiasts.

Some has called blood parrot cichlid "the Frankenstein monster of the fish world." Another notable hybrid, the flowerhorn cichlid, was very popular in some parts of Asia from 2001 until late 2003 and is believed to bring good luck to its owner. The popularity of the flowerhorn cichlid declined in 2004, resulting in many flowerhorn cichlids being released into the rivers and canals of Malaysia and Singapore where they pose a threat to endemic animal communities.

Numerous cichlid species have also been the subject of selective breeding programmes to develop new ornamental strains for the aquarium trade. The most intensive selective breeding programmes have involved angelfish and discuss many mutations that effect both colouration and finnage. Many other cichlids have been selectively bred for albino, leucistic and xanthistic pigment mutations including oscars, convicts and *Pelvicachromis pulcher*. Both dominant and recessively inherited pigment mutations have been observed for cichlids. In convict cichlids, for example, a leucistic colouration is recessively inherited, while in *Oreochromis niloticus niloticus* red colouration is caused by a dominantly inherited mutation.

These efforts at selectively breeding may, however, have unintended consequences. For example, some selectively bred strains of *Mikrogeophagus ramirezi* have health and fertility problems. Similarly, the inbreeding involved in selective breeding programmes can cause severe physical abnormalies such as the notched phenotype in angelfish.

Genera

As of 2006, there are 221 genera:

- *Acarichthys* (Eigenmann 1912)
- *Acaronia* (Myers 1940)
- *Aequidens* (Eigenmann & Bray 1894)
- *Alticorpus* (Stauffer & McKaye 1988)
- *Altolamprologus* (Poll 1986)
- *Amphilophus* (Agassiz, 1859)
- *Anomalochromis* (Greenwood 1985)
- *Apistogramma* (Regan 1913)
- *Apistogrammoides* (Meinken 1965)
- *Archocentrus* (Gill 1877)
- *Aristochromis* (Trewavas, 1935)
- *Astatoreochromis* (Pellegrin 1904)
- *Astatotilapia* (Pellegrin 1904)

- *Astronotus* (Swainson 1839)
- *Aulonocara* (Regan 1922)
- *Aulonocranus* (Regan 1920)
- *Australoheros* (Rican & Kullander 2006)
- *Baileychromis* (Poll 1986)
- *Bathybates* (Boulenger 1898)
- *Benitochromis* (Lamboj 2001)
- *Benthochromis* (Poll 1986)
- *Biotodoma* (Eigenmann & Kennedy 1903)
- *Biotoecus* (Eigenmann & Kennedy 1903)
- *Boulengerochromis* (Pellegrin 1904)
- *Buccochromis* (Eccles & Trewavas 1989)
- *Bujurquina* (Kullander 1986)
- *Callochromis* (Regan 1920)
- *Caprichromis* (Eccles & Trewavas 1989)
- *Caquetaia* (Fowler 1945)
- *Cardiopharynx* (Poll 1942)
- *Chaetobranchopsis* (Steindachner, 1875)
- *Chaetobranchus* (Heckel, 1840)
- *Chalinochromis* (Poll 1974)
- *Champsochromis* (Boulenger 1915)
- *Cheilochromis* (Eccles & Trewavas 1989)
- *Chetia* (Trewavas 1961)
- *Chilochromis* (Boulenger 1902)
- *Chilotilapia* (Boulenger 1908)
- *Chromidotilapia* (Boulenger 1898)
- *Cichla* (Bloch & Schneider 1801)
- *Cichlasoma* (Swainson 1839)
- *Cleithracara* (Kullander & Nijssen 1989)
- *Copadichromis* (Eccles & Trewavas 1989)

- *Corematodus* (Boulenger 1897)
- *Crenicara* (Steindachner 1875)
- *Crenicichla* (Heckel 1840)
- *Ctenochromis* (Pfeffer 1893)
- *Ctenopharynx* (Eccles & Trewavas 1989)
- *Cunningtonia* (Boulenger 1906)
- *Cyathochromis* (Trewavas 1935)
- *Cyathopharynx* (Regan 1920)
- *Cyclopharynx* (Poll 1948)
- *Cynotilapia* (Regan 1922)
- *Cyphotilapia* (Regan 1920)
- *Cyprichromis* (Scheuermann 1977)
- *Cyrtocara* (Boulenger 1902)
- *Danakilia* (Thys van den Audenaerde 1969)
- *Dicrossus* (Steindachner 1875)
- *Dimidiochromis* (Eccles & Trewavas 1989)
- *Diplotaxodon* (Trewavas 1935)
- *Divandu* (Lamboj & Snoeks 2000)
- *Docimodus* (Boulenger 1897)
- *Eclectochromis* (Eccles & Trewavas 1989)
- *Ectodus* (Boulenger 1898)
- *Eretmodus* (Boulenger 1898)
- *Etia* (Schliewen & Stiassny 2003)
- *Etroplus* (Cuvier 1830)
- *Exochochromis* (Eccles & Trewavas 1989)
- *Fossorochromis* (Eccles & Trewavas 1989)
- *Genyochromis* (Trewavas 1935)
- *Geophagus* (Heckel 1840)
- *Gephyrochromis* (Boulenger 1901)
- *Gnathochromis* (Poll 1981)

- *Gobiocichla* (Kanazawa 1951)
- *Grammatotria* (Boulenger 1899)
- *Greenwoodochromis* (Poll 1983)
- *Guianacara* (Kullander & Nijssen 1989)
- *Gymnogeophagus* (Miranda Ribeiro 1918)
- *Haplochromis* (Hilgendorf 1888)
- *Haplotaxodon* (Boulenger 1906)
- *Hemibates* (Regan 1920)
- *Hemichromis* (Peters 1857)
- *Hemitaeniochromis* (Eccles & Trewavas 1989)
- *Hemitilapia* (Boulenger 1902)
- *Herichthys* (Baird & Girard 1854)
- *Heros* (Heckel, 1840)
- *Herotilapia* (Pellegrin 1904)
- *Heterochromis* (Regan 1922)
- *Hoplarchus* (Kaup, 1860)
- *Hoplotilapia* (Hilgendorf 1888)
- *Hypselecara* (Kullander 1986)
- *Hypsophrys* (Agassiz 1859)
- *Interochromis* (Yamaoka, Hori & Kuwamura 1988)
- *Iodotropheus* (Oliver & Loiselle 1972)
- *Iranocichla* (Coad 1982)
- *Ivanacara* (Romer & Hahn 2006)
- *Julidochromis* (Boulenger 1898)
- *Katria* (Stiassny & Sparks 2006)
- *Konia* (Trewavas 1972)
- *Krobia* (Kullander & Nijssen 1989)
- *Labeotropheus* (Ahl 1926)
- *Labidochromis* (Trewavas 1935)
- *Labrochromis* (Greenwood 1980)

- *Laetacara* (Kullander 1986)
- *Lamprologus* (Schilthuis 1891)
- *Lepidiolamprologus* (Pellegrin 1904)
- *Lestradea* (Poll 1943)
- *Lethrinops* (Regan 1922)
- *Lichnochromis* (Trewavas 1935)
- *Limbochromis* (Greenwood 1987)
- *Limnochromis* (Regan 1920)
- *Limnotilapia* (Regan 1920)
- *Lipochromis* (Greenwood 1980)
- *Lithochromis* (Lippitsch & Seehausen 1998)
- *Lobochilotes* (Boulenger 1915)
- *Macropleurodus* (Regan 1922)
- *Maylandia* (Meyer & Foerster 1984)
- *Mazarunia* (Kullander 1990)
- *Mbipia* (Lippitsch & Seehausen 1998)
- *Mchenga* (Stauffer & Konings, 2006)
- *Melanochromis* (Trewavas, 1935)
- *Mesonauta* (Günther, 1862)
- *Microchromis* (Johnson 1975)
- *Mikrogeophagus* (Meulengracht-Madson 1968)
- *Myaka* (Trewavas 1972)
- *Mylacochromis* (Greenwood 1980)
- *Mylochromis* (Regan 1920)
- *Naevochromis* (Eccles & Trewavas 1989)
- *Nandopsis* (Gill, 1862)
- *Nannacara* (Regan 1905)
- *Nanochromis* (Pellegrin 1904)
- *Neetroplus* (Günther, 1867)
- *Neochromis* (Regan 1920)

- *Neolamprologus* (Colombe & Allgayer 1985)
- *Nimbochromis* (Eccles & Trewavas 1989)
- *Nyassachromis* (Eccles & Trewavas 1989)
- *Ophthalmotilapia* (Pellegrin 1904)
- *Oreochromis* (Günther 1889)
- *Orthochromis* (Greenwood 1954)
- *Otopharynx* (Regan 1920)
- *Oxylapia* (Kiener & Mauge 1966)
- *Pallidochromis* (Turner 1994)
- *Parachromis* (Agassiz 1859)
- *Paracyprichromis* (Poll 1986)
- *Paralabidochromis* (Greenwood 1956)
- *Parananochromis* (Greenwood 1987)

Cisco

The ciscoes are salmonid fish of the genus *Coregonus* that differ from other members of the genus in having upper and lower jaws of approximately equal length and high gillraker counts. These species have been the focus of much study recently, as researchers have sought to determine the relationships among species that appear to have evolved very recently.

The five endemic cisco species of the Laurentian Great Lakes- kiyi, bloater, shortnose cisco, longjaw cisco and deep-water cisco are believed to be of very recent evolutionary origin and are said to represent an incipient species flock by some authors. The debate about which were true species and which were not has been rendered academic by the fact that some have disappeared from the system and are apparently extinct.

Several species belong to what various authors refer to as the "*Coregonus artedi*" complex, a group of closely related forms that appear to be derived from the northern cisco (*C. artedi*). The relationships among these forms remains unclear and have complicated discussion of the conservation status of some species.

Ciscoes have been exploited in commercial fisheries, particularly in the Laurentian Great Lakes where the deep-water forms were the basis of the so-called chub fishery. The chub fishery had nothing to do with the various Cyprinid fish species known as chubs but was exclusively based on the various species of ciscos.

The fishery continued as cisco stocks fell and non-native species such as sea lamprey, rainbow smelt and alewife spread through the system and increased in abundance. Alewife, in particular, have been implicated as a predator of cisco eggs and larvae, and as a competitor with ciscos. The fishery shifted focus from species to species as cisco numbers declined and has been largely defunct for some years.

Climbing Gourami

The Anabantidae are a family of perciform fish commonly called the climbing gouramies or climbing perches. As labyrinth fishes, they possess a labyrinth organ, a structure in the fish's head which allows them to "breathe" atmospheric oxygen. Fish of this family are commonly seen gulping the air at the surface of the water; which then passes out of their gills or mouth when they dive beneath the surface.

The climbing gouramies originate from Africa to India and the Philippines. They are primarily a freshwater fish and only very rarely found in brackish water. An egg-layer, they typically guard their eggs and young.

Climbing Gouramis are so names due to their ability to "climb" out of water and "walk" short distances. Their method of terrestrial locomotion uses the gill plates as supports and the fish pushs itself using the fins and tail.

Species

There are about 36 species in the family, classified in four genera (three according to some authors). The familiar aquarium resident, the Siamese Fighting Fish *Betta splendens* used to be classified in this family, but is now placed among the gouramies, family Osphronemidae.

- Genus *Anabas*
  - Gangetic koi, *Anabas cobojius* (Hamilton, 1822).
  - Climbing perch, *Anabas testudineus* (Bloch, 1792).
- Genus *Ctenopoma*
  - Spotted ctenopoma, *Ctenopoma acutirostre* (Pellegrin, 1899).
  - Silverbelly ctenopoma, *Ctenopoma argentoventer* (Ahl, 1922).
  - *Ctenopoma ashbysmithi* (Banister & Bailey, 1979).
  - *Ctenopoma breviventrale* (Pellegrin, 1938).
  - *Ctenopoma ctenotis* (Boulenger, 1920).
  - *Ctenopoma garuanum* (Ahl, 1927).
  - Tailspot ctenopoma, *Ctenopoma kingsleyae* (Günther, 1896).
  - *Ctenopoma machadoi* (Fowler, 1930).
  - *Ctenopoma maculatum* (Thominot, 1886).
  - Manyspined ctenopoma, *Ctenopoma multispine* (Peters, 1844).
  - Ocellated labyrinth fish, *Ctenopoma muriei* (Boulenger, 1906).
  - *Ctenopoma nebulosum* (Norris & Teugels, 1990).
  - Twospot climbing perch, *Ctenopoma nigropannosum* Reichenow, 1875.
  - Eyespot ctenopoma, *Ctenopoma ocellatum* (Pellegrin, 1899).
  - Mottled ctenopoma, *Ctenopoma oxyrhynchum* (Boulenger, 1902).
  - *Ctenopoma pellegrini* (Boulenger, 1902).
  - *Ctenopoma petherici* (Günther, 1864).
  - *Ctenopoma riggenbachi* (Ahl, 1927).
  - *Ctenopoma togoensis* (Ahl, 1928).
  - *Ctenopoma weeksii* (Boulenger, 1896).

- Genus *Microctenopoma*
  - — Ornate ctenopoma, *Microctenopoma ansorgii* (Boulenger, 1912).
  - — Congo ctenopoma, *Microctenopoma congicum* (Boulenger, 1887).
  - — *Microctenopoma damasi* (Poll & Damas, 1939).
  - — Banded ctenopoma, *Microctenopoma fasciolatum* (Boulenger, 1899).
  - — *Microctenopoma intermedium* (Pellegrin, 1920).
  - — *Microctenopoma lineatum* (Nichols, 1923).
  - — *Microctenopoma milleri* (Norris & Douglas, 1991).
  - — Dwarf ctenopoma, *Microctenopoma nanum* (Günther, 1896).
  - — *Microctenopoma nigricans* (Norris, 1995).
  - — *Microctenopoma ocellifer* (Nichols, 1928).
  - — *Microctenopoma pekkolai* (RendAhl, 1935).
  - — *Microctenopoma uelense* (Norris & Douglas, 1995).
- Genus *Sandelia*
  - — Eastern Cape Rocky, *Sandelia bainsii* (Castelnau, 1861).

Cape kurper, *Sandelia capensis* (Cuvier, 1829).

Gobiesocidae

Clingfishes are fishes of the family Gobiesocidae. Most species are marine, being found in shallow waters of the Atlantic, Pacific and Indian Oceans.

They are bottom-dwelling fishes; some species shelter in sea urchins or crinoids. In most species the pelvic fins are modified into a sucking disc.

Classification

The classification of the clingfishes varies. FishBase places Gobiescoidae as the only family in the order Gobiesociformes; Some other classifications, for example ITIS, place them in the suborder Gobiesocoidei of the order Perciformes.

Species

FishBase lists about 152 species in 45 genera:

- Genus *Acyrtops*
  - — Flarenostril clingfish, *Acyrtops amplicirrus* (Briggs, 1955).
  - — Emerald clingfish, *Acyrtops beryllinus* (Hildebrand & Ginsburg, 1926).
- Genus *Acyrtus*
  - — Papillate clingfish, *Acyrtus artius* (Briggs, 1955).
  - — *Acyrtus pauciradiatus* Sampaio, de Anchieta, Nunes & (Mendes, 2004).
  - — Red clingfish, *Acyrtus rubiginosus* (Poey, 1868).
- Genus *Alabes*
  - — *Alabes bathys* (Hutchins, 2006).
  - — *Alabes brevis* (Springer & Fraser, 1976).
  - — Common shore eel, *Alabes dorsalis* (Richardson, 1845).
  - — *Alabes elongata* (Hutchins & Morrison, 2004).
  - — *Alabes gibbosa* (Hutchins & Morrison, 2004).
  - — Dwarf shore eel, *Alabes hoesei* (Springer & Fraser, 1976).
  - — *Alabes obtusirostris* (Hutchins & Morrison, 2004).
  - — *Alabes occidentalis* (Hutchins & Morrison, 2004).
  - — Pygmy shore eel, *Alabes parvulus* (McCulloch, 1909).
  - — *Alabes scotti* (Hutchins & Morrison, 2004).
  - — *Alabes springeri* (Hutchins, 2006).
- Genus *Apletodon*
  - — *Apletodon dentatus bacescui* (Murgoci, 1940).
  - — Small-headed clingfish, *Apletodon dentatus dentatus* (Facciolà, 1887).

— *Apletodon incognitus* (Hofrichter & Patzner, 1997).
— *Apletodon microcephalus* (Brook, 1890).
— Chubby clingfish, *Apletodon pellegrini* (Chabanaud, 1925).

- Genus *Arcos*

— Elegant clingfish, *Arcos decoris* (Briggs, 1969).
— Rockwall clingfish, *Arcos erythrops* (Jordan & Gilbert, 1882).
— Padded clingfish, *Arcos macrophthalmus* (Günther, 1861).
— Galapagos clingfish, *Arcos poecilophthalmos* (Jenyns, 1842).
— Rock clingfish, *Arcos rhodospilus* (Günther, 1864).

- Genus *Aspasma*

— *Aspasma minima* (Döderlein, 1887).

- Genus *Aspasmichthys*

— *Aspasmichthys ciconiae* (Jordan & Fowler, 1902).

- Genus *Aspasmodes*

— *Aspasmodes briggsi* (Smith, 1957).

- Genus *Aspasmogaster*

— Eastern clingfish, *Aspasmogaster costata* (Ogilby, 1885).
— Smooth-snout clingfish, *Aspasmogaster liorhyncha* (Briggs, 1955).
— *Aspasmogaster occidentalis* (Hutchins, 1984).
— Tasmanian clingfish, *Aspasmogaster tasmaniensis* (Günther, 1861).

- Genus *Chorisochismus*

— Rocksucker, *Chorisochismus dentex* (Pallas, 1769).

- Genus *Cochleoceps*

— *Cochleoceps bassensis* (Hutchins, 1983).
— Western cleaner-clingfish, *Cochleoceps bicolor* (Hutchins, 1991).

- Eastern cleaner-clingfish, *Cochleoceps orientalis* (Hutchins, 1991).
- *Cochleoceps spatula* (Günther, 1861).
- Green clingfish, *Cochleoceps viridis* (Hutchins, 1991).

• Genus *Conidens*
- *Conidens laticephalus* (Tanaka, 1909).
- *Conidens samoensis* (Steindachner, 1906).

• Genus *Creocele*
- Broad clingfish, *Creocele cardinalis* (Ramsay, 1883).

• Genus *Dellichthys*
- New Zealand urchin clingfish, *Dellichthys morelandi* (Briggs, 1955).

• Genus *Derilissus*
- *Derilissus altifrons* (Smith-Vaniz, 1971).
- Whiskereye clingfish, *Derilissus kremnobates* (Fraser, 1970).
- *Derilissus nanus* (Briggs, 1969).
- *Derilissus vittiger* (Fraser, 1970).

• Genus *Diademichthys*
- Urchin clingfish, *Diademichthys lineatus* (Sauvage, 1883).

• Genus *Diplecogaster*
- Two-spotted clingfish, *Diplecogaster bimaculata bimaculata* (Bonnaterre, 1788).
- *Diplecogaster bimaculata euxinica* (Murgoci, 1964).
- *Diplecogaster bimaculata pectoralis* (Briggs, 1955).
- *Diplecogaster ctenocrypta* (Briggs, 1955).
- Bigeye clingfish, *Diplecogaster megalops* (Briggs, 1955).

• Genus *Diplocrepis*
- Orange clingfish, *Diplocrepis puniceus* (Richardson, 1846).

• Genus *Discotrema*
- Crinoid clingfish, *Discotrema crinophila* (Briggs, 1976).

- Genus *Ecklonìaichthys*
  - — Weedsucker, *Eckloniaichthys scylliorhiniceps* (Smith, 1943).
- Genus *Gastrocyathus*
  - — New Zealand slender clingfish, *Gastrocyathus gracilis* (Briggs, 1955).
- Genus *Gastrocymba*
  - — *Gastrocymba quadriradiata* (Rendahl, 1926).
- Genus *Gastroscyphus*
  - — Hector's clingfish, *Gastroscyphus hectoris* (Günther, 1876).
- Genus *Gobiesox*
  - — Panamic clingfish, *Gobiesox adustus* (Jordan & Gilbert, 1882).
  - — Clarion clingfish, *Gobiesox aethus* (Briggs, 1951).
  - — Lappetlip clingfish, *Gobiesox barbatulus* (Starks, 1913).
  - — Socorro clingfish, *Gobiesox canidens* (Briggs, 1951).
  - — *Gobiesox crassicorpus* (Briggs, 1951).
  - — *Gobiesox daedaleus* (Briggs, 1951).
  - — Lined clingfish, *Gobiesox eugrammus* (Briggs, 1955).
  - — Mountain clingfish, *Gobiesox fluviatilis* (Briggs & Miller, 1960).
  - — *Gobiesox fulvus* (Meek, 1907).
  - — Peninsular clingfish, *Gobiesox juniperoserrai* (Espinosa Perez & Castro-Aguirre, 1996).
  - — *Gobiesox juradoensis* (Fowler, 1944).
  - — Bahama skilletfish, *Gobiesox lucayanus* (Briggs, 1963).
  - — Northern clingfish, *Gobiesox maeandricus* (Girard, 1858).
  - — Lonely clingfish, *Gobiesox marijeanae* (Briggs, 1960).
  - — *Gobiesox marmoratus* (Jenyns, 1842).
  - — Mexican clingfish, *Gobiesox mexicanus* (Briggs & Miller, 1960).
  - — *Gobiesox milleri* (Briggs, 1955).

— *Gobiesox multitentaculus* (Briggs, 1951).
— *Gobiesox nigripinnis* (Peters, 1860).
— *Gobiesox nudus* (Linnaeus, 1758).
— Bearded clingfish, *Gobiesox papillifer* (Gilbert, 1890).
— *Gobiesox pinniger* (Gilbert, 1890).
— *Gobiesox potamius* (Briggs, 1955).
— Stippled clingfish, *Gobiesox punctulatus* (Poey, 1876).
— California clingfish, *Gobiesox rhessodon* (Smith, 1881).
— Smoothlip clingfish, *Gobiesox schultzi* (Briggs, 1951).
— *Gobiesox stenocephalus* (Briggs, 1955).
— Skilletfish, *Gobiesox strumosus* (Cope, 1870).
— Woods' clingfish, *Gobiesox woodsi* (Schultz, 1944).

- Genus *Gouania*
  — Blunt-snouted clingfish, *Gouania willdenowi* (Risso, 1810).
- Genus *Gymnoscyphus*
  — *Gymnoscyphus ascitus* (Böhlke & Robins, 1970).
- Genus *Haplocylix*
  — Giant clingfish, *Haplocylix littoreus* (Forster, 1801).
- Genus *Kopua*
  — Kuiter's deepsea clingfish, *Kopua kuiteri* (Hutchins, 1991).
  — *Kopua nuimata* (Hardy, 1984).
- Genus *Lecanogaster*
  — *Lecanogaster chrysea* (Briggs, 1957).
- Genus *Lepadichthys*
  — Bolin's clingfish, *Lepadichthys bolini* (Briggs, 1962).
  — Pale clingfish, *Lepadichthys caritus* (Briggs, 1969).
  — Eyestripe clingfish, *Lepadichthys coccinotaenia* (Regan, 1921).
  — *Lepadichthys ctenion* (Briggs & Link, 1963).
  — *Lepadichthys erythraeus* (Briggs & Link, 1963).

— *Lepadichthys frenatus* (Waite, 1904).
— Doubleline clingfish, *Lepadichthys lineatus* (Sauvage, 1883).
— Minor clingfish, *Lepadichthys minor* (Briggs, 1955).
— *Lepadichthys sandaracatus* (Whitley, 1943).
— *Lepadichthys springeri* (Briggs, 2001).

• Genus *Lepadogaster*
— Connemarra clingfish, *Lepadogaster candolii* (Risso, 1810).
— Shore clingfish, *Lepadogaster lepadogaster* (Bônnaterre, 1788).
— Cornish sucker, *Lepadogaster purpurea* (Bonnaterre, 1788).
— *Lepadogaster zebrina* (Lowe, 1839).

• Genus *Liobranchia*
— Minute clingfish, *Liobranchia stria* (Briggs, 1955).

• Genus *Lissonanchus*
— Streaky clingfish, *Lissonanchus lusheri* (Smith, 1966).

• Genus *Modicus*
— *Modicus minimus* (Döderlein, 1887).
— *Modicus tangaroa* (Hardy, 1983).

• Genus *Opeatogenys*
— *Opeatogenys cadenati* (Briggs, 1957).
— *Opeatogenys gracilis* (Briggs, 1955).

• Genus *Parvicrepis*
— Little clingfish, *Parvicrepis parvipinnis* (Waite, 1906).

• Genus *Pherallodichthys*
— *Pherallodichthys meshimaensis* (Shiogaki & Dotsu, 1983).

• Genus *Pherallodiscus*
— Northern fraildisc clingfish, *Pherallodiscus funebris* (Gilbert, 1890).
— Southern fraildisc clingfish, *Pherallodiscus varius* (Briggs, 1955).

- Genus *Pherallodus*
  - *Pherallodus indicus* (Weber, 1913).
  - Mini-clingfish, *Pherallodus smithi* (Briggs, 1955).
- Genus *Posidonichthys*
  - *Posidonichthys hutchinsi* (Briggs, 1993).
- Genus *Propherallodus*
  - *Propherallodus briggsi* (Smith, 1957).
- Genus *Rimicola*
  - Channel Islands clingfish, *Rimicola cabrilloi* (Briggs, 2002).
  - Southern clingfish, *Rimicola dimorpha* (Briggs, 1955).
  - Slender clingfish, *Rimicola eigenmanni* (Gilbert, 1890).
  - Kelp clingfish, *Rimicola muscarum* (Meek & Pierson, 1895).
  - Guadalupe clingfish, *Rimicola sila* (Briggs, 1955).
- Genus *Sicyases*
  - *Sicyases brevirostris* (Guichenot, 1848).
  - *Sicyases hildebrandi* (Schultz, 1944).
  - *Sicyases sanguineus* (Müller & Troschel, 1843).
- Genus *Tomicodon*
  - Distant clingfish, *Tomicodon absitus* (Briggs, 1955).
  - *Tomicodon abuelorum* (Szelistowski, 1990).
  - *Tomicodon australis* (Briggs, 1955).
  - Bifid clingfish, *Tomicodon bidens* (Briggs, 1969).
  - Cortez clingfish, *Tomicodon boehlkei* (Briggs, 1955).
  - *Tomicodon briggsi* (Smith, 1957).
  - Smallsucker clingfish, *Tomicodon chilensis* (Brisout de Barneville, 1846).
  - *Tomicodon clarkei* (Williams & Tyler, 2003).
  - *Tomicodon cryptus* (Williams & Tyler, 2003).
  - Rosy clingfish, *Tomicodon eos* (Jordan & Gilbert, 1882).
  - Barred clingfish, *Tomicodon fasciatus* (Castelnau, 1861).

— Sonora clingfish, *Tomicodon humeralis* (Gilbert, 1890).
— *Tomicodon lavettsmithi* (Williams & Tyler, 2003).
— *Tomicodon leurodiscus* (Williams & Tyler, 2003).
— Blackstripe clingfish, *Tomicodon myersi* (Briggs, 1955).
— Peter's clingfish, *Tomicodon petersii* (Garman, 1875).
— *Tomicodon prodomus* (Briggs, 1969).
— *Tomicodon reitzae* (Briggs, 2001).
— *Tomicodon rhabdotus* (Smith-Vaniz, 1969).
— *Tomicodon rupestris* (Poey, 1860).
— *Tomicodon vermiculatus* (Briggs, 1955).
— Zebra clingfish, *Tomicodon zebra* (Jordan & Gilbert, 1882).

- Genus *Trachelochismus*
  — Striped clingfish, *Trachelochismus melobesia* (Phillipps, 1927).
  — New Zealand lumpfish, *Trachelochismus pinnulatus* (Forster, 1801).

Clown Loach

The clown loach *Chromobotia macracanthus*, (formerly *Botia macracanthus*) is a freshwater fish belonging to the loach family (Cobitidae) and is the sole member of the genus *Chromobotia*. Originating in Asia on the islands of Sumatra and Borneo, the species is popular in the aquarium trade.

Distribution and Habitat

Clown loaches are restricted to the Indonesian islands of Sumatra and Borneo and are found in warm (25-30°C; 77-86°F), fast-moving, acidic water (pH < 7.0).

Appearance

They have orange fins with 3 black bands on an orange body and some barbels about their mouth which are used to feel around for food. Often sold in the pet trade at around 1.5 inches, they do grow fast until about 5 inches at which time their growth slows to a maximum length of 11.8 in (30.0 cm)

Although they are unlikely to reach this size in a tank, those from Borneo tend to have a black area in the pelvic fins, while in Sumatran fish the fin is completely orange. A movable spine rests in a groove below the eye, which should be treated with caution when netting the fish.

Diet: They eat small worms, crustaceans and snails.

Reproduction and Sexual Dimorphism

There are no known cases of clowns breeding in captivity, partly due to the fact that they do not reach sexual maturity until after nine or ten years of age. Clown loaches may live for up to 50 years.

In the Aquarium

A harmless, very active, social fish, they are best kept in groups of 5 or more and due to their potential size an aquarium of 6 feet x 2 feet x 2 feet should be the minimum size used. These fish have bifurcated subocular (located under the eyes) spines, which are used as a defence mechanism. If a loach deploys its spines while caught in a net, untangling it is difficult and can cause severe injury. It is also a good idea when moving larger specimens to double or triple bag them or use a solid container. Some owners have been stabbed while trying to catch or touch these fish. When kept in groups smaller than 5, they may spend more time hiding under obstacles in the water.

They are also found to make clicking noises when excited or eating. Sometimes they lie on their sides on the bottom of the tank and appear to be dead. This is a common event and the aquarist should be aware of this fact or unnecessary removal may occur.

A good tank setup for a clown loach should include ample shade, plants (plastic or real), hiding places and other peaceful fishes. Make sure the environment is not too bright. Provide shade from tank lighting. The tank should not be next to a window. Clown loaches are keen observers of other fish in the aquarium; they observe and react accordingly. If other fish are skittish and hide, clowns will observe this and do the same. Make sure that other fish in your community tank are docile and not prone to hide.

Because clown loaches come from rivers and streams, they are accustomed to having other fishes and plants in their environment. Not having plants and/or other fishes can cause clown loaches to become stressed and to go into hiding. Another important thing to remember is that since they do come from a fast moving river environment, they need a tank with lots of clear, well filtered and fast moving water which can be achieved with proper filtration and the use of powerheads. Before introducing clown loaches to your tank, make sure the fish you currently have are compatible because aggressive fish will stress your clown loaches and may need to be removed. Also, clown loaches are particularly susceptible to Ichthyophthirius (ich), so watch them closely when initially adding them to your tank and when you introduce new fish. Ich usually deals the typical clown loach a poor prognosis since the standard treatment is especially toxic to the clown's "skin-type" and the dose must be halved and is therefore less effective.

You should also provide a variety of foods for your loaches. Upon feeding your clown loaches, observe their behaviour. In a community tank always make sure that enough food gets to the bottom where these fish usually feed. Most clown loaches accept commercial flake food as their dietary staple, but thrive with a variety of food: live (worms, brine shrimp, small snails), plant matter, freeze-dried (tubifex worms, especially if it is fortified) and frozen brine shrimp (always thaw frozen food to aquarium temperature).

Clown loaches are also regarded as a natural way of controlling an infestation of small snails in the aquarium. This being said, a person considering them for this purpose must also consider their future needs with regards to a large aquarium. A person getting clowns to removed snails in a smaller tank might be better served with one of the many other botia species that are as effective at the task but remain much smaller. Despite that utilitarian purpose, clown loaches are usually kept for reasons of appearance and personality.

Clown Triggerfish

The clown triggerfish, *Balistoides conspicillum*, is a triggerfish from the order Tetraodontiformes. This reef-associated fish is commonly found in the tropical Indo-Pacific and Red Sea.

Distribution and Habitat: This species is a primarily marine species. This fish is found in Tropical Indo-Pacific and Red Sea coastal waters from 1-75 metres in depth (3-250 ft). This fish is generally uncommon or rare throughout its range, which includes East Africa to South Africa, through to Indonesia, and all the way to Japan and New Caledonia.

The clown triggerfish is most commonly found around coral reefs. It lives in clear coastal to outer reef habitats. It also occurs in clear, seaward reefs near steep drop-offs.

Anatomy and Appearance: The fish can reach up to about 50 cm (20 in) in length. It has strong jaws which can be used to crush and eat sea urchins, crustaceans and hard-shelled molluscs.

This fish has unique colouration. The ventral surface has large, white spots on a dark background, and its dorsal surface has black spots on yellow. There is a vertical, white (slightly yellow) stripe on the caudal fin. The brightly painted yellow mouth may be used to deter potential predators. This fish has a form of camouflage that is, or is similar to, countershading. From below, the white spots look like the surface of the water above it. From above, the fish will blend in more with the coral reef environment.

Ecology

This fish is a solitary species. Adults are often seen near steep drop-offs swimming openly, though may hide when approached. Juveniles are more secretive and occur in small caves with rich invertebrate growth. These fish eat sea urchins, crabs and other crustaceans, molluscs, and tunicates.

In the Aquarium

Because of its attractive colouration, this fish is one of the most highly prized aquarium fish. Like many other triggerfish, it can require a large aquarium and be aggressive towards other fish. It should not be kept with small fishes. It will also prey on invertebrates in the aquarium. This fish can become tame enough to be hand-fed, however one should beware of the fish's sharp teeth.

Cobbler

The cobbler, *Cnidoglanis macrocephalus,* is an eeltail catfish found along the coasts of Australia, favouring shallow protected waters near river mouths. It is also known as deteira, estuary catfish, and South Australian catfish.

Like other eeltail catfish, the cobbler resembles a catfish in front, but an eel behind. Its mottled green body ranges up to 91 cm in length. The dorsal and pectoral fins have spines. Cobblers stay in holes and under ledges during the day, then come out at night to feed on molluscs, crustaceans, polychaete worms, algae, and organic debris.

The South Australian cobbler, *Gymnapistes marmoratus,* is also called just "cobbler" in Australia; the term is also used in the Caribbean for several species in *Trachinotus.*

Cobia

'Cobia (Rachycentron canadum) *- also known as black kingfish, black salmon, ling, lemonfish, crabeaters, aruan tasek,* etc.—are perciform marine fish, the sole representative of their family, the Rachycentridae.

Description: Attaining a length of 2 metres (78 inches) and a weight of 68 kilograms (150 pounds), Cobia have elongate fusiform (spindle shaped) bodies and broad, flattened heads. Their eyes are small and their lower jaw projects slightly past the upper jaw.

On the jaws, tongue and roof of the mouth are bands of *villiform* (fibrous) teeth. Their bodies are smooth with small scales, their dark brown colouration grading to white on the belly with two darker brown horizontal bands on the flanks. These may not be prominent except during spawning when Cobia lighten in colour and adopt a more prominently striped pattern. The large pelvic fins are normally carried horizontally (rather than vertically as shown for convenience in the illustration), so that, as seen in the water they may be mistaken for a small shark.

When boated, the horizontal pelvic fins enable the Cobia to remain upright so that their vigorous thrashing can make them a hazard. The first dorsal fin is composed of six to nine ,independent, short, stout, and sharp spines. The family name *Rachycentridae,* from the Greek words *rhachis* meaning "spine" and *kentron* meaning "sting," is an allusion to these dorsal spines. Mature cobia have forked, slightly lunate tail fins with most fins being a dark brown. They lack air bladders.

Similar Species: Cobia somewhat resemble and are most closely related to the Remora of the family Echeneidae. However, they lack the dorsal sucker of the Remora, their body is far stouter and their tail is far more developed, and forked instead of rounded. Juvenile Cobia are patterned with conspicuous bands of black and white. Their tails are rounded rather than forked as in the adults.

Distribution and Habitat

Cobia are pelagic and are normally solitary except for annual spawning aggregations; however they will congregate at reefs, wrecks, harbours, buoys and other structural oases. They may also enter estuaries and mangroves in search of prey.

Feeding Habits

Cobia feed primarily on Crabs, squid and other fish. Cobia will follow larger animals such as Sharks, Turtles and Manta Rays in hope of scavenging a meal.

Cobia are intensely curious fish and show no fear of boats. Their predators are not well documented, but the dolphinfish (*Coryphaena hippurus*) is known to feed on immature Cobia. Shortfin mako sharks are known to feed on adult Cobia and have been seen by fishermen following Cobia during their annual springtime migration in the northern Gulf of Mexico.

Life History

Cobia are pelagic spawners; that is, they release many tiny (1.2 mm) buoyant eggs into the water which become part of the plankton. The eggs float freely with the currents until hatching. The larvae are also planktonic, being more or less helpless during their first week until the eyes and mouths develop. Males mature at two years and females at three years. Both sexes lead moderately long lives of 15 years or more. Spawning takes place diurnally from April to September in large offshore congregations. Up to 20 individual spawns may take place in one season, with intervals of about one to two weeks.

Cobia are frequently parasitised by nematodes, trematodes, cestodes, copepods and acanthocephalans.

Migration Patterns

Cobia make seasonal migrations along the coasts in search of water in their preferred temperature range. Wintering in the Gulf of Mexico, they migrate north as far as Maryland in the Summer, passing East Central Florida in March.

Fishing Gear and Methods

Cobia are powerful fish popular among sport fishermen, and a prized table fare. Cobia are primarily sought by sight casting small Chartreuse jigs to migrating fish, or to fish following Turtles or Manta rays. They are a bycatch taken by commercial fishermen trolling for King Mackerel, or by live Menhaden slow-trolled near reefs and wrecks.

Utilisation

Cobia are sold commercially, and command a high price for their firm texture and excellent flavour. However, there is no directed fishery owing to their solitary nature. They have been farmed in aquaculture for this reason. The meat is usually sold fresh. They are typically served in the form of grilled or poached fillets.

Cod

Cod is the common name for the genus *Gadus* of fish, belonging to the family Gadidae, and is also used in the common name of a variety of other fishes. Cod is a popular food fish with a mild flavour, low fat content and a dense white flesh that flakes easily. Cod livers are processed to make cod liver oil, an important source of Vitamin A, Vitamin D and omega-3 fatty acids (EPA and DHA). Also the cod fish can change colours at certain depths of the water.

The Atlantic cod has two distinct colour phases, gray-green and reddish brown. Its average weight is 10 to 25 lb (4.5—11.3 kg), but specimens weighing up to 200 lb (90 kg) have been recorded. Young Atlantic cod or haddock prepared in strips for cooking is called scrod. Cods feed on molluscs, crabs, starfish, worms, squid, and small fish. Some migrate south in winter to spawn. A large female lays up to five million eggs in midocean, a very small number of which survive. The Pacific cod is found N of Oregon. The tomcod resembles a young Atlantic cod with long, tapering ventral fins. It rarely exceeds 15 in. (37.5 cm) in length and lives close to shore. There is also a Pacific tomcod. The pollack, also called coalfish or green cod, is a plump olive-green cod found in cool waters of the Atlantic. Pollacks have forked tails and pale lateral lines and grow to 3 ft (90 cm) and 30 lb (13.6 kg). Some grow to 6 feet in length.

In the United Kingdom, Atlantic cod is one of the most

common kinds of fish to be found in fish and chips, along with haddock and plaice.

It is also well known for being largely consumed in Portugal, where it is considered a treasure of the nation's cuisine. It is an important link in the food chain.

Species in Genus Gadus

At various times in the past, a very considerable number of species have been classified in this genus. However the great majority of them are now either classified in other genera, or have been recognised as simply forms of one of three species. Modern taxonomy, therefore, recognises only three species in this genus:

- Atlantic cod, *Gadus morhua*
- Pacific cod, *Gadus macrocephalus*
- Greenland cod, *Gadus ogac*

All these species have a profusion of common names, most of them including the word "cod". Many common names have been used of more than one species, in different places or at different times.

Related Species Called Cod

Cod forms part of the common name of many other fish no longer classified in the genus *Gadus*. Many of these are members of the family Gadidae, and several were formerly classified in genus *Gadus;* others are members of three related families whose names include the word "cod": the morid cods, Moridae (100 or so species); the eel cods, Muraenolepididae (4 species); and the Eucla cod, Euclichthyidae (1 species). The tadpole cod family (Ranicipitidae) has now been absorbed within Gadidae.

Species within the order Gadiformes that are commonly called cod include:

- Arctic cod *Arctogadus glacialis*
- East Siberian cod *Arctogadus borisovi*
- Saffron cod *Eleginus gracilis*

- Polar cod *Boreogadus saida*
- Rock cod *Lotella rhacina*
- Poor cod *Trisopterus minutus*
- Pelagic cod *Melanonus gracilis*
- Small-headed cod *Lepidion microcephalus*
- Tadpole cod *Guttigadus globosus*
- Eucla cod *Euclichthys polynemus*

Some other related fish have common names derived from "cod", such as codling, codlet or tomcod. ("Codling" is also used as a name for a young cod.)

Unrelated Species Called Cod

However there are also fish commonly known as cod that are quite unrelated to the genus *Gadus*. Part of this confusion of names is market-driven. Since the decline in cod stocks has made the Atlantic cod harder to catch, cod replacements are marketed under names of the form "*x* cod", and culinary rather than phyletic similarity has governed the emergence of these names. A very large number of fish have thus been named as some kind of cod at some time. The following species, however, seem to have well established common names including the word "cod"; note that all are Southern Hemisphere species.

Perciformes

Fish of the order Perciformes that are commonly called "cod" include:

- Murray cod *Maccullochella peelii peelii*
- Eastern freshwater cod *Maccullochella ikei*
- Mary River cod *Maccullochella peelii mariensis*
- Trout cod *Maccullochella macquariensis*
- Sleepy cod *Oxyeleotris lineolatus*
- Blue cod *Parapercis colias*
- The cod icefish family, Nototheniidae, including:

— Black cod *Paranotothenia microlepidota*

— Maori cod *Paranotothenia magellanica*

— Antarctic cod *Dissostichus mawsoni*

Rock Cod, Reef Cod, and Coral Cod

Almost all the fish known as coral cod, reef cod or rock cod are also in order Perciformes. Most are better known as groupers, and belong to the family Serranidae. Others belong to the Nototheniidae. Two exceptions are the Australasian red rock cod, which belongs to a different order, and the fish known simply as the rock cod also by soft cod in New Zealand, *Lotella rhacina*, is related to the true cod (it is a morid cod).

Scorpaeniformes

From the order Scorpaeniformes:

- Ling cod *Ophiodon elongatus*
- Red rock cod *Scorpaena papillosa*

Ophidiiformes

The tadpole cod family, Ranicipitidae, and the Eucla cod family, Euclichthyidae, were formerly classified in the order Ophidiiformes, but are now grouped with the Gadiformes.

Species Marketed as Cod

Some fish that do not have "cod" in their names are sometimes sold as cod. Haddock and whiting belong in the same family, the Gadidae, as cod.

- Haddock *Melanogrammus aeglefinus*
- Whiting *Merlangius merlangus*

Identification

Classic codfish shape, with three rounded dorsal and two anal fins. The pelvic fins are small with the first ray extended, and are set under the gill cover (i.e. the throat region), in front of the pectorals. The upper jaw extends over the lower jaw, which has a well developed chin barbel. Medium sized eyes, approximately the same as the length of the chin barbel.

It has a distinct white lateral line running from the gill slit above the pectoral fin, to the base of the caudal or tail fin. The back tends to be a greenish to sandy brown, and showing extensive mottling especially towards the lighter sides and white belly. Dark brown colouration of the back and sides is not uncommon especially for individuals who have resided in rocky inshore regions.

Breeding

The Cod population comprises of a number of reasonably distinct stocks over its range. These include the Arcto-Norwegian, North Sea, Faroe, Iceland, East Greenland, West Greenland, Newfoundland, and Labrador stocks. There would seem to be little interchange between the stocks, although migrations to their individual breeding grounds may involve distances of 200 miles or more.

Spawning occurs between January to April (March and April are the peak months), at a depth of 200m in specific spawning grounds at water temperatures of between 4-6°C. Around the UK, the major ones are associated with the Middle to Southern North Sea, the start of the Bristol Channel (north of Newquay), the Irish Channel (both east and west of the Isle of Man), around Stornoway, and east of Helmsdale.

Pre-spawning courtship involves fin displays, and male grunting, which leads to pairing. The male is inverted underneath the female, whilst the pair swim in circles during the spawning process.

The eggs are planktonic and hatch between 8 to 23 days with the larva being some 4mm in length. This planktonic phase lasts some ten weeks, during which the young cod will increase it's body weight by 40 times, and be about 2cm in length.

The young cod move to the seabed and their diet changes to small benthic crustaceans, such as isopods and small crabs. They increase in size to 8cm in the first six months, 14 to 18cm by the end of their first year, and some 25 to 35cm by the end of the second. This rate of growth tends to be less in individuals occupying northerly grounds. Cod reach maturity at about 50cm in length at about 3 to 4 years of age.

Habitat

Varied, although often favouring rough ground especially inshore. Demersal in depths of between 20 to 200m (80m Av.), although not uncommon to depths of 600m. Gregarious and forms schools, although shoaling tends to be a feature of the spawning season.

Food

Adult cod are active hunters, feeding on sandeels, whiting, haddock, small cod, and squid, crabs, mussels, worms, mackerel, and molluscs supplementing their diets. While young cod eat the same they may not eat food sources that are too large.

Range

Throughout most of the UK waters, although generally seen as a winter fish in the south.

Cod Trade

Cod has been an important economic commodity in an international market since the Viking period (around 800 AD). Norwegians used dried cod during their travels and soon a dried cod market developed in southern Europe. This market has lasted for more than 1000 years, passing through periods of Black Death, wars and other crises and still is an important Norwegian fish trade.

The Portuguese since the 15th century have been fishing cod in the North Atlantic and clipfish is widely eaten and appreciated in Portugal. The Basques also played an important role in the cod trade.

Apart from the long history this particular trade also differs from most other trade of fish by the location of the fishing grounds, far from large populations and without any domestic market.

The large cod fisheries along the coast of North Norway (and in particular close to the Lofoten islands) have been developed almost uniquely for export, depending on sea transport of stockfish over large distances. Since the introduction of salt, dried salt cod ('klippfisk' in Norwegian) has also been exported. The trade operations and the sea transport were by the end of the 14th century taken over by the Hanseatic League, Bergen being the most important port of trade.

William Pitt the Elder, criticising the Treaty of Paris in Parliament, claimed that cod was British gold; and that it was folly to restore Newfoundland fishing rights to the French.

In the 17th and 18th centuries, the New World, especially in Massachusetts and Newfoundland, cod became a major commodity, forming triangular trade networks and cross-cultural exchanges.

In the 20th century, Iceland re-emerged as a fishing power and entered the Cod Wars to gain control over the north Atlantic seas. In the late 20th and early 21st centuries, cod fishing off the coast of Europe and America severely depleted cod stocks there which has since become a major political issue as the necessity of restricting catches to allow fish populations to recover has run up against opposition from the fishing industry and politicians reluctant to approve any measures that will result in job losses. The 2006 Northwest Atlantic cod quota is set at 23,000 tons representing half the available stocks, while it is set to 473,000 tons for the Northeast Atlantic cod.

Nototheniidae

The cod icefishes are a family, Nototheniidae, of Perciform fishes, containing about 50 species in 12 genera. They are largely found in the Southern Ocean and off the coast of Antarctica. As the dominant Antarctic fish taxa, they occupy both sea bottom and water column ecological niches. Although lacking a gas bladder, they have undergone a depth-related diversification, such as increase in fatty tissues and reduced mineralisation of the bones, resulting in a body density approaching neutral, to fill a variety of water column niches. It is thought that the spleen may be used to remove ice crystals from circulating blood. As the chilly subantarctic waters averages –1 to 4 degrees Celsius, most Antarctic species have antifreeze glycoproteins in their blood and other body fluids.

Some species exhibit polymorphism, for example, the circum-Antarctic *Trematomus newnesi* exists as two morphs in the Ross Sea, the typical morph and a large mouth/broad headed morph.

As the major fish resource in the Southern Ocean, nototheniids are under increasing pressure from commercial fishing.

Classification

- Genus *Aethotaxis*
  - — *Aethotaxis mitopteryx*
- Genus *Cryothenia*
  - — *Cryothenia amphitreta*
  - — *Cryothenia peninsulae*
- Genus *Dissostichus*
  - — Patagonian toothfish, *Dissostichus eleginoides*
  - — Antarctic cod, *Dissostichus mawsoni*
- Genus *Gobionotothen*
  - — *Gobionotothen acuta*
  - — *Gobionotothen barsukovi*
  - — *Gobionotothen gibberifrons*
  - — *Gobionotothen marionensis*
- Genus *Gvozdarus*
  - — *Gvozdarus svetovidovi*
- Genus *Lepidonotothen*
  - — *Lepidonotothen kempi*
  - — *Lepidonotothen larseni*
  - — *Lepidonotothen macrophthalma*
  - — *Lepidonotothen mizops*

- — *Lepidonotothen nudifrons*
- — *Lepidonotothen squamifrons*

- Genus *Notothenia*
  - — Maori chief, *Notothenia angustata* (Hutton, 1875)
  - — *Notothenia coriiceps*
  - — *Notothenia cyanobrancha*
  - — Black cod or smallscaled cod, *Notothenia microlepidota* (Hutton, 1875)
  - — Marbled rockcod, *Notothenia rossii*
- Genus *Pagothenia*
  - — *Pagothenia borchgrevinki*
  - — *Pagothenia brachysoma*
- Genus *Paranotothenia*
  - — *Paranotothenia dewitti*
  - — Maori cod, Maori chief, black cod, Magellanic rockcod, blue notothenia or orange throat notothen, *Paranotothenia magellanica* (Forster, 1801)
- Genus *Patagonotothen* (all non-Antarctic)
  - — *Patagonotothen brevicauda*
  - — *Patagonotothen canina*
  - — *Patagonotothen cornucola*
  - — *Patagonotothen elegans*
  - — *Patagonotothen guntheri*
  - — *Patagonotothen jordani*
  - — *Patagonotothen kreffti*
  - — *Patagonotothen longipes*
  - — *Patagonotothen ramsayi*
  - — *Patagonotothen sima*
  - — *Patagonotothen squamiceps*
  - — *Patagonotothen tessellata*
  - — *Patagonotothen thompsoni*
  - — *Patagonotothen wiltoni*

- Genus *Pleuragramma*
  - — Antarctic silverfish, *Pleuragramma antarcticum*
- Genus *Trematomus*
  - — *Trematomus bernacchii*
  - — *Trematomus eulepidotus*
  - — *Trematomus hansoni*
  - — *Trematomus lepidorhinus*
  - — *Trematomus loennbergii*
  - — *Trematomus newnesi*
  - — *Trematomus nicolai*
  - — *Trematomus pennellii*
  - — *Trematomus scotti*
  - — *Trematomus tokarevi*
  - — *Trematomus vicarius*

Codlet

Codlets are a family, Bregmacerotidae, of cod-like fishes, containing the single genus *Bregmaceros*.

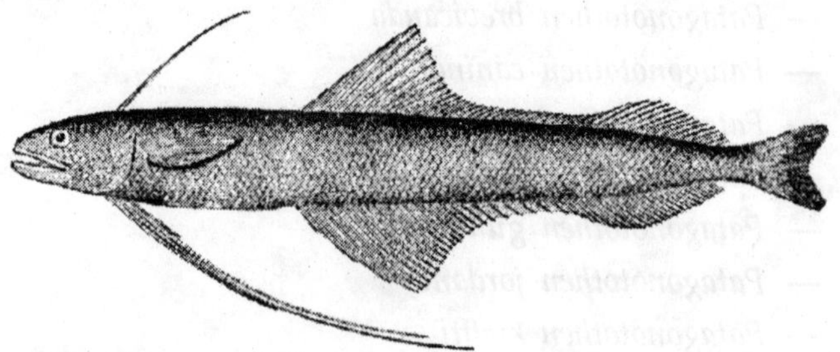

They are found in tropical and subtropical waters throughout the world. They are very small fishes: the big-eye unicorn-cod, *Bregmaceros rarisquamosus*, reaches no more than 1.5 cm, and even the largest, *Bregmaceros lanceolatus*, reaches only 11.5 cm.

Their scientific name is from Greek *bregma* meaning the top of the head, and *keras* meaning "horn"; this refers to their occipital ray (a spine emerging from the top of the head).

Species

There are twelve species:

- *Bregmaceros arabicus* (D'Ancona & Cavinato, 1965).
- Antenna codlet, *Bregmaceros atlanticus* (Goode & Bean, 1886).
- Codlet, *Bregmaceros bathymaster* (Jordan & Bollman, 1890).
- Striped codlet, *Bregmaceros cantori* (Milliken & Houde, 1984).
- Stellate codlet, *Bregmaceros houdei* (Saksena & Richards, 1986).
- Japanese codlet, *Bregmaceros japonicus* (Tanaka, 1908).
- *Bregmaceros lanceolatus* (Shen, 1960).
- Spotted codlet, Macclelland's unicorn-cod, or Unicorn cod, *Bregmaceros mcclellandi* (Thompson, 1840).
- Smallscale codlet, *Bregmaceros nectabanus* (Whitley, 1941).
- *Bregmaceros neonectabanus* (Masuda, Ozawa & Tabeta, 1986).
- *Bregmaceros pescadorus* (Shen, 1960).
- Big-eye unicorn-cod, *Bregmaceros rarisquamosus* (Munro, 1950).

Coelacanth

Coelacanth ('hollow spine' in Greek, *coelia* (*êïeëeÜ*) meaning hollow and *acanthos* (*Üêáíeïò*) spine) IPA: [EsiÐlYÌkæne] is the common name for an order of fish that includes the oldest living lineage of jawed fish known to date. The coelacanths, which are closely related to lungfishes, were believed to have been extinct since the end of the Cretaceous period, until a live specimen was found off the east coast of South Africa, off the Chalumna River in 1938. Since then, they have been found in the Comoros, Sulawesi (Indonesia), Kenya, Tanzania, Mozambique, Madagascar, Greater St. Lucia Wetland Park in South Africa, and more recently, sister-species in Sulawesi (Indonesia), considerably increasing the geographical distribution ascribed to this species.

Biological Characteristics

Coelacanths first appear in the fossil record in the Middle Devonian, about 410 million years ago. Prehistoric species of coelacanth lived in many bodies of water in Late Paleozoic and Mesozoic times.

The average weight of the living west Indian Ocean coelacanth, *Latimeria chalumnae*, is 176 pounds (80kg) and they can reach up to 6.5 feet (2m) in length. Adult females are slightly larger than males. Based on growth rings in their ear bones (otoliths) scientists infer that individual coelacanths may live as long as 80 to 100 years. Coelacanths live as deep as 700m (2300ft) below sea level, but are more usually found at depths of 90 to 200m.

Living examples of *Latimeria chalumnae* have a deep blue colour which probably camouflages them from prey species, however the Indonesian species is brown. *Latimeria chalumnae* is widely but very sparsely distributed around the rim of the west Indian Ocean, seemingly occurring in small colonies. Coelacanth eyes are very sensitive, and have a *tapetum lucidum*. Coelacanths are almost never caught in the daytime or on nights with full moons, due to the sensitivity of their eyes. Coelacanth eyes also have many rods: tiny structures that help animals see in dim light. Together, the rods and tapetum help the fish see better in dark water.

Coelacanths are the only living species known to have a functional intracranial joint, which almost completely separates the front and back halves of the skull internally. Flexure at this joint may aid in the consumption of large prey by the use of suction. Coelacanths are also mucilaginous; their scales release mucus and their bodies continually exude oil. This oil is a laxative, and makes the fish virtually inedible unless dried and salted. Although most fish have smooth scales, only the lower half of the coelacanth's scales are smooth. Toothlike spines called denticles cover the upper half of each scale, making it rough and scratchy. The smooth lower half of each scale is protected by the denticles of the two scales that overlap it. These rough, layered scales provide armourlike protection against predators and the rough edges of rocks. The rough scales are used by the natives of Comoros as sandpaper.

Coelacanths are opportunistic feeders, hunting cuttlefish, squid, snipe eels, small sharks, and other fish found in their deep reef and volcanic slope habitats. Coelacanths are also known to swim head down, backwards and belly up to locate their prey presumably utilising its rostral gland. Scientists suspect that one reason this fish has been so successful is that they can slow down their metabolisms at any time, sinking into the less-inhabited depths and minimising their nutritional requirements in a sort of hibernation mode.

The coelacanths which live near Sodwana Bay, South Africa, rest in caves at depths of 90 to 150m during daylight hours, but disperse and swim to depths as shallow as 55m when hunting at night.

The depth is not as important as their need for very dim light, and water which has a temperature of 14 to 22°C, and they will rise or sink to find these conditions. (Ref: Dr. Anthony Ribbink, coelacanth expert.) Coelacanths are lobe-finned fish with the pectoral and anal fins on fleshy stalks supported by bones, and the tail or caudal fin diphycercal (divided into three lobes), the middle one of which also includes a continuation of the notochord. Coelacanths have modified cosmoid scales, which are thinner than true cosmoid scales, which can only be found on extinct fish. Coelacanths also have a special electroreceptive device called a rostral organ in the front of the skull, which probably helps in prey detection.

Fossil Record

Although now represented by only two living species, as a group the coelacanths were once very successful with many genera and species that left an abundant fossil record from the Devonian to the end of the Cretaceous period, at which point they apparently suffered a nearly complete extinction, and past which point no fossils are known. It is often claimed that the coelacanth has remained unchanged for millions of years but in fact the living species and even genus are unknown from the fossil record. However, some of the extinct species, particularly those of the last known fossil coelacanth, the

Cretaceous genus *Macropoma*, closely resemble the living species. The most likely reason for the gap is the taxon having become extinct in shallow waters. Deep water fossils are only rarely lifted to levels where paleontologists can recover them, making most deep water taxa disappear from the fossil record. This situation is still under investigation by scientists.

Reproduction

Females give birth to between 5 and 25 young, which are capable of surviving on their own immediately after birth. Coelacanths are ovoviviparous. Coelacanths give birth to live young called pups. Their reproductive behaviours are not well known, but it is believed that they are not sexually mature until after 20 years of age. Gestation time is 13 months.

Discovery

Timeline of Discoveries Date	*Description*
1938	(December 23) Discovery of the first modern coelacanth 30km SW of East London, South Africa.
1952	(December 21) Second specimen identified in the Comoros. Since then more than 200 have been caught around the islands.
1988	First photographs of coelacanths in their natural habitat, by Hans Fricke off Grand Comore.
1991	First coelacanth identified near Mozambique, 24km offshore NE of Quelimane.
1995	First recorded coelacanth on Madagascar, 30km S of Tulear.
1997	(September 18) New species of coelacanth found in Indonesia.
2000	A group found by divers off Sodwana Bay, South Africa.
2001	A group found off the coast of Kenya.
2003	First coelacanth caught by fisherman in Tanzania. Within the year, 22 were caught in total.
2004	Canadian researcher William Sommers captured the largest recorded specimen of coelacanth off the coast of Madagascar.
2007	(May 19) Indonesian fisherman Justinus Lahama caught a 4 feet long, 112 pound coelacanth off Sulawesi Island near Bunaken National Marine Park that survived for 17 hours in a quarantined pool.

First Find in South Africa

On December 23, 1938, Hendrik Goosen, the captain of the trawler *Nerine* returned to the harbour at East London after a trawl around the mouth of the Chalumna River. As he frequently did, he telephoned his friend, Marjorie Courtenay-Latimer, curator at East London's small museum to see if she wanted to look over the contents of the catch for anything interesting. At the harbour Latimer noticed a blue fin and took a closer look. There she found what she later described as "the most beautiful fish I had ever seen, five feet long, and a pale mauve blue with iridescent silver markings."

Failing to find a description of the creature in any of her books, she attempted to contact her friend, Professor James Leonard Brierley Smith, but he was away for Christmas.

Unable to preserve the fish, she reluctantly sent it to a taxidermist. When Smith returned, he immediately recognised it as a coelacanth, known only from fossils. Smith named the fish *Latimeria chalumnae* in honour of Marjorie Courtenay-Latimer and the waters in which it was found. The two discoverers received immediate recognition and the fish became known as a "living fossil." The 1938 coelacanth is still on display in the East London Museum.

However, as the specimen had been stuffed, the gills or skeleton were not available for examination and some doubt did remain as to whether it was truly the same species. Smith began a hunt for a second specimen that would take more than a decade.

Comoros

A worldwide search was launched for more coelacanths, with a reward of 100 British pounds, a very substantial sum to the average South African fisherman of the time. Fourteen years later, one specimen was found in the Comoros, but the fish was no stranger to the locals—in the port of Mutsamudu on the Comorian island of Anjouap, the Comorians were puzzled to be so rewarded for a *gombessa* or *mame*, an inferior, nearly inedible fish that their fishermen occasionally caught by mistake.

The second specimen, found in 1952 by Comorian fisherman Ahamadi Abdallah, was described as a different species, first as 'Malania hunti' and later as *Malania anjounae,* after Daniel François Malan, the South African Prime Minister who had dispatched an SAAF Dakota at the behest of Professor Smith to fetch the specimen.

It was later discovered that the lack of a first dorsal fin, at first thought to be significant, was caused by an injury early in the specimen's life. Ironically, Malan was a staunch creationist; startled by the sight of the ancient creature, he exclaimed, "Why it's ugly! Is this where we come from?" The specimen retrieved by Smith is on display at the SAIAB in Grahamstown, South Africa where Smith worked.

The Comorians are now aware of the significance of the endangered species and have established a programme to return accidentally caught coelacanth to deep water.

As for Smith, who died in 1968, his account of the coelacanth story appeared in the book *Old Fourlegs,* first published in 1956. His book *Sea Fishes of the Indian Ocean,* illustrated and co-authored by his wife Margaret, remains the standard ichthyological reference for the region.

In 1988, National Geographic photographer, Hans Fricke was the first to photograph the species in its natural habitat, 180 m (590 ft 7 in) off Grand Comore's west coast.

Second Species in Indonesia

On September 18, 1997, Arnaz and Mark Erdmann, travelling in Indonesia on their honeymoon, saw a strange fish enter the market at Manado Tua, on the island of Sulawesi. A second specimen was preserved on July 30, 1998. Arnaz thought it was a gombessa, although it was brown, not blue. An expert noticed their pictures on the Internet and realised its significance. DNA testing revealed that this species, called *raja laut* ("King of the Sea") by the Indonesians, is not related to the Comorian population. This fish was described in a 1999 issue of *Environmental Biology of Fishes* by Pouyaud *et al.* It was given the scientific name *Latimeria menadoensis*.

A recent molecular study estimated the divergence time between the two coelacanth species to be 40-30 mya. On May 19, 2007, an Indonesian fisherman caught a 110 pound coelacanth off Sulawesi island near Bunaken National Marine Park. He shortly kept the "living fossil" in a separate pool. It died 17 hours later which was unexpectedly long, because these creatures live in cold water. The local university is now studying the carcass.

St. Lucia Marine Protected Area in South Africa

In South Africa, the search continued on and off over the years. 46-year-old diver Riaan Bouwer lost his life searching for coelacanths in June 1998.

On the 28th of October 2000, just south of the Mozambique border in Sodwana Bay in the St. Lucia Marine Protected Area, three deep-water divers, Pieter Venter, Peter Timm, and Etienne le Roux, made a dive to 104 metres and unexpectedly spotted a coelacanth.

Calling themselves "SA Coelacanth Expedition 2000", the group returned with photographic equipment and several additional members. On the 27th of November, after an unsuccessful initial dive the previous day, four members of the group, Pieter Venter, Gilbert Gunn, Christo Serfontein, and Dennis Harding, found three coelacanths. The largest was between 1.5 and 1.8 metres in length; the other two were from 1 to 1.2 metres. The fish swam head-down and appeared to be feeding from the cavern ledges. The group returned with video footage and photographs of the coelacanths.

During the dive, however, Serfontein lost consciousness, and 34-year-old Dennis Harding rose to the surface with him in an uncontrolled ascent. Harding complained of neck pains and died from a cerebral embolism while on the boat. Serfontein recovered after being taken underwater for decompression sickness treatment.

In March–April of 2002, the Jago Submersible and Fricke Dive Team descended into the depths off Sodwana and observed fifteen coelacanths. A dart probe was used to collect tissue samples. The shallowest recorded sighting of a coelacanth is at a depth of 58m off the coast of Sodwana Bay by Christo Vanjaarsveld.

Taxonomy

Subclass Coelacanthimorpha (Actinistia) are sometimes used to designate the group of Sarcopterygian fish that contains the Coelacanthiformes. The following is a classification of known coelacanth genera and families:

Class Sarcopterygii

Subclass Coelacanthimorpha

- Order Coelacanthiformes
  - — Family Coelacanthidae (extinct)
    - ⇒ *Axelia* (extinct)
    - ⇒ *Coelacanthus* (extinct)
    - ⇒ *Ticinepomis* (extinct)
    - ⇒ *Wimania* (extinct)
  - — Family Diplocercidae (extinct)
    - ⇒ *Diplocercides* (extinct)
  - — Family Hadronectoridae (extinct)
    - ⇒ *Allenypterus* (extinct)
    - ⇒ *Hadronector* (extinct)
    - ⇒ *Polyosteorhynchus* (extinct)
  - — Family Mawsoniidae (extinct)
    - ⇒ *Alcoveria* (extinct)
    - ⇒ *Axelrodichthys* (extinct)
    - ⇒ *Chinlea* (extinct)
    - ⇒ *Diplurus* (extinct)
    - ⇒ *Holophagus* (extinct)
    - ⇒ *Mawsonia* (extinct)

— Family Miguashaiidae (extinct)
  ⇒ *Miguashaia* (extinct)
— Family Latimeriidae
  ⇒ *Holophagus* (extinct)
  ⇒ *Libys* (extinct)
  ⇒ *Macropoma* (extinct)
  ⇒ *Macropomoides* (extinct)
  ⇒ *Megacoelacanthus* (extinct)
  ⇒ *Latimeria* (James Leonard Brierley Smith, 1939)
  ⇒ *L. chalumnae* (Comorese coelacanth) (James Leonard Brierley Smith, 1939)
  ⇒ *L. menadoensis* (Indonesian coelacanth) (Pouyaud, Wirjoatmodjo, Rachmatika, Tjakrawidjaja, *et al.*, 1999)
  ⇒ *Undina* (extinct)
— Family Laugiidae (extinct)
  ⇒ *Coccoderma* (extinct)
  ⇒ *Laugia* (extinct)
— Family Rhabdodermatidae (extinct)
  ⇒ *Caridosuctor* (extinct)
  ⇒ *Rhabdoderma* (extinct)
— Family Whiteiidae (extinct)
  ⇒ *Whiteia* (extinct)

Coho Salmon

The Coho salmon is a species of anadromous fish in the salmon family. Coho salmon are also known as silver salmon or "silvers".

During their ocean phase, Coho have silver sides and dark blue backs. During their spawning phase, the jaws and teeth of the coho become hooked, and they develop bright red sides, bluish green heads and backs, dark bellies with dark spots on their back.

Sexually maturing coho develop a light pink or rose shading along the belly and the males may show a slight arching of the back.

Mature coho salmon have a pronounced red skin colour with darker backs and average 38 inches in length and seven to 11 pounds in weight, although coho weighing up to 36 pounds have been reported. Mature females may be darker than males, with both showing a pronounced hook on the nose.

The eggs hatch in the spring. The young spend one to two years in the fresh water before migrating to the ocean in late March through July. Young often spend the first winter in off-channel sloughs. Some fish leave fresh water in the spring, spend summer in brackish estuarine ponds and then migrate back into fresh water in the fall. Coho salmon live in the salt water for one or two years before returning to spawn.

The traditional range of the coho salmon runs from both sides of the North Pacific ocean, from Hokkaidô, Japan and eastern Russian, around the Bering Sea to mainland Alaska, and south all the way to Monterey Bay, California. Coho salmon have also been introduced in Lake Erie, as well as many other landlocked reservoirs throughout the United States.

Coho salmon are the backbone of the Alaska troll fishery, however, the majority are caught by the net fishery (Gillnet and Seine). Coho salmon average 3.5 per cent by fish of the annual Alaska salmon harvest; 5.9 per cent by weight of the annual Alaska salmon harvest.

This species is a game fish and provides fine sport in fresh and salt water from July to December, especially with light tackle. It is one of the most popular sport fish in the Pacific Northwest of the United States. Its popularity is due in part to the reckless abandon which it frequently displays chasing baits and lures while in salt water, and the large number of coastal streams it ascends during its spawning runs. Its habit of schooling in relatively shallow water, and often near beaches, makes it accessible to anglers on the banks as well as in boats.

Ocean caught coho is regarded as excellent table fare. It has a moderate to high amount of fat, which is considered essential when judging taste. Only Spring Chinook and Sockeye salmon have higher levels of fats in their meat.

On May 6, 1997, The National Marine Fisheries Service, on behalf of the Secretary of Commerce, listed as threatened the Southern Oregon/Northern California Coast "Evolutionarily Significant Unit" of coho salmon. The coho salmon population in the southern Oregon/Northern California region has declined from an estimated 150,000 to 400,000 naturally spawning fish in the 1940s to less than 10,000 naturally producing adults today. The dramatic reduction in the coho salmon population has been due to many natural and man-made conditions, including long-term trends in atmospheric conditions, such as El Nino, which causes extremes in annual rainfall on the northern California coast, the predation of coho salmon by California sea lions and Pacific harbour seals, and commercial timber harvesting.

Historically, the coho, along with other species, has been a staple in the diet of several Native American tribes, who would also use it to trade with other tribes farther inland.

The coho salmon is also a symbol of several Native American tribes, representing life and sustenance.

Collared Carpetshark

The collared carpetshark, *Parascyllium collare,* is a carpetshark of the family Parascylliidae found off eastern Australia, between latitudes 26° S and 38° S, at depths of between 20 and 160 m. Its length is up to 85 cm.

The collared carpetshark has a caudal fin with its upper lobe not elevated above the body axis, with a strong terminal lobe and subterminal notch but no ventral lobe. It is a common but little-known shark found on the continental shelf.

Colouration is light yellowish to reddish-brown, with a prominent dark brownish collar around the gills, five dusky saddles, and dark spots on the body, tail and fins. Reproduction is oviparous, with flattened, elongate egg-cases.

Colorado Pikeminnow

The Colorado pikeminnow (formerly squawfish) *Ptychocheilus lucius* is the largest cyprinid fish of North America, with reports of individuals up to 6 ft long and weighing over 100 lb. Formerly an important food fish for both Native Americans and European settlers, and widespread in the Colorado River basin of the southwestern United States, its numbers and range have declined in historical times, and it is now listed as an endangered species.

Like the other pikeminnows, it has an elongated body reminiscent of the pike. The cone-shaped and somewhat flattened head is elongated, nearly a quarter of the body length. Colour grades from bright olive green on the back to a paler yellowish shade on the sides, to white underneath. Young fish also have a dark spot on the caudal fin. Both the dorsal and anal fins typically have nine rays. The pharyngeal teeth are long and hooked.

The reports of 6-ft individuals are estimates from skeletal remains, although a number of "old-timers" interviewed by *Salt Lake Tribune* in 1994 reported that such individuals were common. Catches in the 1960s ranged up to 60 cm for 11-year-old fish, although in the early 1990s maximum sizes reached no more than 34 cm. Biologists now consider the average size of an adult Pikeminnow to be between 4 and 9 pounds, with reports of the fish commonly exceeding 3 feet in length now in question.

The pikeminnow's diet almost entirely consists of fish. Young individuals, up to 5 cm long, eat cladocerans, copepods, and chironomid larvae, then shift to insects at around 10 cm, gradually eating more fish as they mature.

The pikeminnow was never a popular species of sportfish or food fish, and once introduced species, like bass, catfish, and pike became more numerous in the rivers, locals were more than happy to make the switch.

Their usual habitat is the backwaters of the turbulent and turbid rivers that make up the Colorado system.

Although once found from Wyoming to Mexico, damming and habitat eliminations has reduced the species to the upper Colorado drainage; populations are known from the Green River, Gunnison River, White River, San Juan River, and Yampa River. They have been transplanted to the Salt River and Verde River, both outside their native range.

Efforts to recover the Colorado Pikeminnow have thus far failed and the fish continue to decline, with the species being extirpated from most areas it once lived in.

Combfish

The combfish or comb wrasse, *Coris picta*, is a wrasse of the genus *Coris*, found off eastern Australia and around offshore islands off north eastern New Zealand to depths of between 5 and 60 metres, on mixed sandy/rocky reef areas. Its length is between 10 and 25 centimetres.

The combfish has a long white body with a prominent wide black stripe from its mouth through its eye along the body to the end of the tail. The lower margin of the stripe is wavy and comb-like giving the fish its common name. The portion of the stripe on the tail fin turns yellow during the breeding season. There is also a thin red stripe running along the top of the body from the mouth, along the base of the dorsal fin for the fin's full length.

Combfish often act as cleaners and some get most of their food in this way. The contrasting colour pattern with its cleaning signal stripe acts as a strong attraction to many other reef fish and combfish are often surrounded by groups of fishes waiting to be cleaned. The rest of their diet consists of small crustaceans.

Combtooth Blenny

Combtooth blennies are blennioids; perciform marine fish of the family Blenniidae. They are the largest family of blennies, with approximately 371 species in 53 genera represented. Combtooth blennies are found in tropical and subtropical waters in the Atlantic, Pacific and Indian Oceans; some species are also found in brackish and even freshwater environments.

Physical Description

The body plan of the combtooth blennies is archetypal to all other blennioids; their blunt heads and eyes are large, with large continuous dorsal fins (which may have 3-17 spines). Their bodies are compressed, elongate and scaleless; the small, slender pelvic fins (which are absent in only two species) are situated before the enlarged pectoral fins, and the tail fin is rounded. As their name would suggest, combtooth blennies are noted for their comb-like teeth lining their jaws.

By far the largest species is the eel-like hairtail blenny at 53 centimetres in length; most other members of the family are much smaller. Combtooth blennies are active and often highly colourful, making them popular in the aquarium hobby.

Habitat and Behaviour

Generally benthic fish, combtooth blennies spend much of their time on or near the bottom. They may inhabit the rocky crevices of reefs, burrows in sandy or muddy substrates, or even empty shells.

Generally found in shallow waters, some combtooth blennies are capable of leaving the water for short periods during low tide, aided by their large pectoral fins which act as "feet". Small benthic crustaceans, molluscs and other sessile invertebrates are the primary food items for most species; others eat algae or plankton.

There is one exceptional group of combtooth blennies which deserve special mention: the so-called sabre-toothed blennies of the genera *Aspidontus, Meiacanthus, Petroscirtes, Plagiotremus,* and *Xiphasia*. These blennies have fang-like teeth with venom glands at their bases.

Species of the genera *Apistodontus* and *Plagiotremus* (such as the false cleanerfish) are noted for their cunning mimickry of cleaner wrasses: by imitating the latter's colour, form and behaviour, the blennies are able to trick other fish (or even divers) into letting down their guard, long enough for the blennies to nip a quick mouthful of skin or scale.

Some combtooth blennies will form small groups, while others are solitary and territorial. They may be either diurnal or nocturnal, depending on the species. Females lay eggs in shells or under rock ledges; males guard the nest of eggs until hatching. In some species, the eggs may remain in the oviduct of the female until hatched.

The fry of some species undergo an *ophioblennius* stage, wherein the fish are pelagic (i.e., inhabiting the midwater) and have greatly enlarged pectoral fins and hooked teeth.

Genera

- *Aidablennius*
- *Alloblennius*
- *Alticus*
- *Andamia*
- *Antennablennius*
- *Aspidontus*
- *Atrosalarias*
- *Bathyblennius*
- *Blenniella*
- *Blennius*
- *Chalaroderma*
- *Chasmodes*
- *Cirripectes*
- *Cirrisalarias*
- *Coryphoblennius*
- *Crossosalarias*
- *Dodekablennos*
- *Ecsenius*
- *Enchelyurus*
- *Entomacrodus*
- *Exallias*

- *Glyptoparus*
- *Haptogenys*
- *Hirculops*
- *Hypleurochilus*
- *Hypsoblennius*
- *Istiblennius*
- *Laiphognathus*
- *Lipophrys*
- *Litobranchus*
- *Lupinoblennius*
- *Meiacanthus*
- *Mimoblennius*
- *Nannosalarias*
- *Oman*
- *Omobranchus*
- *Omox*
- *Ophioblennius*
- *Parablennius*
- *Parahypsos*
- *Paralipophrys*
- *Paralticus*
- *Parenchelyurus*
- *Pereulixia*
- *Petroscirtes*
- *Phenablennius*
- *Plagiotremus*
- *Praealticus*
- *Rhabdoblennius*
- *Salaria*
- *Salarias*

- *Scartella*
- *Scartichthys*
- *Spaniblennius*
- *Stanulus*
- *Xiphasia*

Common Carp

The Common carp or European carp (*Cyprinus carpio*) is a widespread freshwater fish distantly related to the common goldfish (*Carassius auratus*), with which it is capable of interbreeding. It gives its name to the carp family Cyprinidae. Common carp are native to Asia and Eastern Europe.

It has been introduced into environments worldwide. It can grow to a maximum length of 5 feet (1.5 metres), a maximum weight of over 80lb (37.3 kg), and an oldest recorded age of at least 65 years. The wild, non-domesticated, forms tend to be much less stocky at around 20 per cent - 33 per cent the maximum size. Koi ((nishikigoi) in Japanese, (pinyin: l- yú) in Chinese) is a domesticated ornamental variety that originated in China but became known to the Western world through Japan.

Although they are very tolerant of most conditions, the common carp prefer large bodies of slow or standing water and soft, vegetative sediments. A schooling fish, they prefer to be in groups of 5 or more. They natively live in a temperate climate in fresh or brackish water with a 7.0 - 7.5 pH, a water hardness of 10.0 - 15.0 dGH, and an ideal temperature range of 37.4 - 75.2°F (3 - 24°C).

Diet

The common carp, as well as its variants, mirror carp, with large mirror like scales (linear mirror - scaleless except for a row of large scales that run along the lateral line; originating in Germany), leather carp (virtually unscaled except near dorsal fin) and fully scaled carp, is omnivorous and will eat almost anything that it comes across. The common carp is happy to eat a vegetarian diet of water plants, but also insects, crustaceans (including zooplankton), or even dead fish if the opportunity arises.

Common Carp as Pests

Common carp have been introduced, often illegally, into many countries. In some countries, due to their habit of grubbing through bottom sediments for food and alteration of their environment, they destroy, uproot and disturb submerged vegetation causing serious damage to native duck and fish populations. In Australia, there is enormous anecdotal and mounting scientific evidence that introduced carp are the cause of permanent turbidity and loss of submergent vegetation in the Murray-Darling river system, with severe consequences for river ecosystems, water quality and native fish species.

Efforts to eradicate a small colony from a Tasmania's Lake Crescent without chemicals have been successful, however the long-term, expensive and intensive undertaking is an example of the both the possibility and difficulty of safely removing the species once it is established.

Common carp have attributes that allow them to be an invasive species - a species that invades and dominates new ecosystems with serious negative effects to the ecosystem and native fauna.

In Victoria (Australia), Common carp has been declared as noxious fish species therefore, there is no restriction on the quantity that a fisher can take. In South Australia, it is an offence for this species to be released back to the wild. An Australian company has made good use of common carp while helping the environment by churning them into plant fertilizer.

Common carp was brought to the US in 1831. In the late 1800s they were distributed widely throughout the USA by the government as a foodfish. However, common carp are not now normally prized as a foodfish in the United States. As in Australia, their introduction has been shown to have negative environmental consequences and they are usually considered to be invasive species. Millions of dollars are spent annually by natural resource agencies to control common carp populations in the United States.

Catching and Eating Carp

Common carp are extremely popular with anglers in many parts of Europe, and their popularity as quarry is slowly increasing among anglers in the United States. Very specialised baits and tackle have been developed for common carp angling.

Carp is also eaten in many parts of the world both when caught from the wild and raised in aquaculture. In the Czech Republic, Slovakia, Croatia and Poland, a carp is a traditional part of a Christmas Eve dinner.

Reproduction

An egg-layer, a typical adult fish can lay 300,000 eggs in a single spawning. Research shows that carp can spawn multiple times in a season in some areas. The young are preyed upon by other predatorial fish such as the northern pike and largemouth bass.

Northern Bluefin Tuna

The northern bluefin tuna (*Thunnus thynnus*) is a species of tuna fish, living in both the Western and the Eastern Atlantic Ocean and extending into the Mediterranean Sea and the Black Sea. Although not native to the Pacific Ocean, it is cultivated off Japan. Northern bluefin tuna can live to be up to 30 years old. The typical size is 2 m (6.6 ft) at about 500 kg (1,100 lb). The largest recorded specimen was caught off Nova Scotia, and was recorded as weighing 679 kg (1,500 lb). They are caught by sports fishermen using a heavy-duty rod and reel. The northern bluefin tuna is an important food fish used almost exclusively in sushi; canned tuna and tuna sold as steaks are of other species.

The species is also known as horse mackerel and in the past was called the common tunny. It is often referred to simply as the "bluefin" or "bluefin tuna", but this name is ambiguous as it is also sometimes used for the southern bluefin tuna, *Thunnus maccoyii*, and the Pacific bluefin tuna, *T. orientalis*. Even the preferred name, northern bluefin tuna, is not unambiguous, because this is sometimes used for the longtail tuna *T. tonggol*. In Australia, canned *T. tonggol* can and is legally sold under the name "Northern bluefin tuna".

The body of the northern bluefin tuna is cigar-shaped and robust. The head is conical and the mouth rather large. The colour is dark blue above and gray below. Northern bluefin tuna can easily be distinguished from other members of the tuna family by the relatively short length of their pectoral fins. Their livers have a unique and definitive characteristic in that they are covered with blood vessels (striated). In other tunas with short pectoral fins, such vessels are either not present or present in small numbers along the edges.

The northern bluefin tuna is an important source of seafood, providing most of the tuna used in sushi. It is a particular delicacy in Japan where the price of a single giant tuna can exceed $100,000 on the Tsukiji fish market in Tokyo. It is also popular in Taiwan, particularly in the town of Donggang. As a result, some fisheries of bluefin are considered overfished, and this problem is compounded by the bluefin's slow growth rate and late maturity. The Atlantic population of the species has declined by nearly 90 per cent since the 1970s. The bluefin species are consequently listed as ones to "Avoid" on the Monterey Bay Aquarium's Seafood Watch programme.

The tetraphyllidean tapeworm *Pelichnibothrium speciosum* has been found to parasitise this species (Scholz *et al.* 1998). As the tapeworm's definite host is the blue shark which does not generally seem to feed on tuna, it is likely that the northern bluefin tuna is a dead-end host for *P. speciosum*.

Conger eel

"Conger" or "conger eel" is a vernacular term used for a number of different species of fish, mostly eels of the family Congridae, and especially the genus *Conger*.

Likely possibilities include:

- *Anguilla reinhardtii*
- *Ariosoma balaricum*
- *Bassanago bulbiceps (Swollen-headed conger eel)*
- *Bathycongrus thysanochilus*
- *Conger cinareus*
- *Conger conger*
- *Conger esculentus*
- *Conger oceanicus*
- *Conger triporiceps*
- *Conger verreauxi*
- *Gnathophis cinctus*
- *Gymnothorax moringa*
- *Hoplunnis tenuis*
- *Paraconger caudilimbatus*
- *Rhynchoconger flavus*
- *Rhynchoconger gracilior*

Over 150 additional common names of fish use "conger" in combination.

Cookiecutter Shark

The cookiecutter shark, *Isistius brasiliensis*, also known as the cigar shark or luminous shark, is a small rarely-seen dogfish shark.

Anatomy and Morphology

The cookiecutter sharks often glow green and grow up to 50 cm (20 in) long. The underside of the shark is bioluminescent, glowing a pale blue-green that matches the background light from the ocean's surface that serves as camouflage to creatures beneath it. However, a small non-luminescent patch appears black, deceiving the shark's prey, smaller predatory fish (like tuna), into thinking the shark is an even smaller fish. When the predatory fish tries to strike at the shark, the shark strikes back, scoring itself another meal.

This is the only known instance whereby a bioluminescent lure is created by the absence of luminescence (contrast with anglerfish).

Distribution

The cookiecutter shark has been found at depths of about 1,000 m (3,300 ft) below the surface of the ocean.

Ecology and Life History

Feeding Ecology: It derived its name from its habit of removing small circular chunks of flesh from whales and large fish. It is hypothesised that the shark seizes its much larger prey with its jaws, then rotates its body to achieve a highly symmetrical cut. They are considered parasites.

Life History

Cookiecutter sharks reproduce through aplacental viviparity in the same way as great white sharks. Little else is known about their reproduction.

Etymology and Taxonomic History

Its name comes from its feeding style which often creates perfect "cookie-cutter" shaped plugs in the skin of large marine mammals and other large sharks.

Interaction with Humans

There has been little interaction between humans and the cookiecutter shark. However, there was an incident in which a cookiecutter shark took a bite out of the rubber sonar dome of a US Navy submarine, causing damage to the housing, and forcing the submarine out of service until the rubber could be replaced.

Kuhli Loach

The kuhli loach, *Pangio kuhlii*, is a small worm-like freshwater fish belonging to the loach family (Cobitidae). It originates in Indonesia and the Malay Peninsula.

Physical Description

The kuhli loach is an eel-shaped fish, elongated with slightly compressed sides, and very small fins. The dorsal fin starts behind the middle of the body, and the anal fin well behind this. The eyes are covered with a transparent skin. The body has 15 to 20 dark brown to black vertical bars, and the gaps between them are salmon pink to yellowish with a light underside. When the fish are not actively breeding, there is no known way to distinguish between males and females. When breeding, the females often become monstrously fat before spawning. Spawning is not easy, but when it occurs a few hundred greenish eggs are laid among the roots of floating plants. Maximum size is 4 in (10 cm) long (matures at 7 cm), and four pairs of barbels about their mouth.

Habitat, Diet and Related Information

The natural habitat of the kuhli loach is the sandy beds of slow-moving rivers and clean mountain streams. They are a social fish and are typically found in small clusters (they are not schooling fish but enjoy the company of their species), but are cautious and nocturnal by nature and swim near the bottom where they feed around obstacles. They natively live in a tropical climate and prefer water with a 5.5 - 6.5 pH, a water hardness of 5.0 dGH, and a temperature range of 75 - 86 °F (24 - 30 °C).

Other Noteworthy Information

In the wild, the fish spawn communally in very shallow water. The kuhli loach is a bottom dweller which burrows into soft places. Its ordinal name 'Acanthophthalmus' comes from the meaning 'thorn' or 'prickle-eye', after a spine beneath each eye.

Etymology of Name

The kuhli loach was originally described as *Cobitis kuhlii* by Achille Valenciennes in 1846. The fish is also commonly called coolie loach, giant coolie loach, slimy loach, and leopard loach. In scientific literature, it has been referred to as *Acanthophthalmus kuhlii*. The genus name *Acanthophthalmus* is a junior synonym of *Pangio*.

In the Aquarium

The kuhli loach is commonly kept as a pet in tropical aquaria. There are a number of species of the genus *Pangio* that appear similar and are sold under the same name, require similar care, and are all excellently suited for household tanks. They tend to be hardy and long-lived in the aquarium and get along well with their own kind as well as others.

In an aquarium environment, especially if the gravel is suitably finely grained, *Pangio* species can burrow into the bottom and there remain unseen for long periods of time, emerging to eat during the night. If the gravel is later disturbed, a hobbyist might well find themselves faced with fish assumed lost a long time ago.

Cornetfish

The cornetfishes are a small family Fistulariidae of extremely elongated fishes in the order Syngnathiformes. The family consists of just a single genus *Fistularia* with four species, found worldwide in tropical and subtropical marine environments.

Ranging up to 200 cm in length, cornetfishes are as thin and elongate as many eels, but are distinguished by a very long snout, distinct dorsal and anal fins, and a forked caudal fin whose centre rays form a lengthy filament. The lateral line is well-developed and extends onto the caudal filament.

They generally live in coastal waters or on coral reefs, where they feed on small fishes, crustaceans and other invertebrates.

Cornetfish are of minor interest for fishing, and can be found in local markets within their range.

Species

- Genus *Fistularia*
  - — Bluespotted cornetfish, *Fistularia commersonii* (Rüppell, 1838)

— Pacific cornetfish, *Fistularia corneta* (Gilbert & Starks, 1904)
— Red cornetfish, *Fistularia petimba* (Lacepede, 1803)
— Cornet fish, *Fistularia tabacaria* (Linnaeus, 1758)

Boxfish

The boxfishes are a family, Ostraciidae, of squared, bony fish belonging to the order Tetraodontiformes, closely related to the pufferfishes and filefishes. They come in a variety of different colours, and are notable for the hexagonal or "honeycomb" patterns in their skin and skeletons. They swim in a rowing manner. Fish in the family are known variously as boxfishes, cofferfishes, cowfishes and trunkfishes.

Boxfishes occupy the Atlantic, Indian, and Pacific oceans, generally at middle latitudes, although the common or buffalo trunkfish (*Lactophrys trigonus*) which lives mainly in Florida waters may be found as far north as Cape Cod. The cowfish variety *Lactophrys quadricornis* can grow to be 50 cm or less in size, but are generally smaller at higher latitudes.

The hexagonal plate-like scales of these fish are fused together into a solid, triangular, box-like carapace, from which the fins and tail protrude. Young boxfishes have a more rounded in shape and may exhibit bright colours. Because of their body scale structure, boxfishes are limited to slow movements.

In the Aquarium

Boxfish generally are very personable fish yet are poorly suited for the home aquaria. They are not thought of as an ordinary aquarium tank mate, but are quickly gaining popularity.

They do pose a hazard in the community tank however. They are capable of releasing a very powerful toxin which can kill other fish and in some cases, the boxfish itself. They generally only use it when threatened or dying, but can become disturbed easily with aggressive tank mates or overcrowded aquarium. Generally boxfish are not reef safe.

Genera and Species

- Genus *Acanthostracion*
  - — *Acanthostracion guineensis* (Bleeker, 1865).
  - — Island cowfish, *Acanthostracion notacanthus* (Bleeker, 1863).
  - — Honeycomb cowfish, *Acanthostracion polygonius* (Poey, 1876).
  - — Scrawled cowfish, *Acanthostracion quadricornis* (Linnaeus, 1758).
- Genus *Lactophrys*
  - — Spotted trunkfish, *Lactophrys bicaudalis* (Linnaeus, 1758).
  - — Buffalo trunkfish, *Lactophrys trigonus* (Linnaeus, 1758).
  - — Smooth trunkfish, *Lactophrys triqueter* (Linnaeus, 1758).
- Genus *Lactoria*
  - — Longhorn cowfish, *Lactoria cornuta* (Linnaeus, 1758).
  - — Roundbelly cowfish, *Lactoria diaphana* (Bloch & Schneider, 1801).
  - — Thornback cowfish, *Lactoria fornasini* (Bianconi, 1846).
  - — *Lactoria paschae* (Rendahl, 1921).
- Genus *Ostracion*
  - — Yellow boxfish, *Ostracion cubicus* (Linnaeus, 1758).
  - — Bluetail trunkfish, *Ostracion cyanurus* (Rüppell, 1828).
  - — *Ostracion immaculatus* (Temminck & Schlegel, 1850).
  - — Whitespotted boxfish, *(Ostracion meleagris* Shaw, 1796).
  - — Shortnose boxfish, *(Ostracion nasus* Bloch, 1785).
  - — Horn-nosed boxfish, *(Ostracion rhinorhynchos* Bleeker, 1852).
  - — Reticulate boxfish, *Ostracion solorensis* (Bleeker, 1853).
  - — Roughskin trunkfish, *Ostracion trachys* (Randall, 1975).
  - — Whitley's box, *Ostracion whitleyi* (Fowler, 1931).
- Genus *Paracanthostracion*
  - — *Paracanthostracion lindsayi* (Phillipps, 1932).

- Genus *Tetrosomus*

— Triangular boxfish, *Tetrosomus concatenatus* (Bloch, 1785).

— Humpback turretfish, *Tetrosomus gibbosus* (Linnaeus, 1758).

— Smallspine turretfish, *Tetrosomus reipublicae* (Ogilby, 1913).

— *Tetrosomus stellifer* (Bloch & Schneider, 1801).

Cownose Ray

The cownose ray (*Rhinoptera bonasus*) is the most common type of ray found in the Chesapeake Bay. The rays grow rapidly and male rays are about 35 inches (900 mm) in width and weigh 26 pounds (12 kg). Females are 28 inches (700 mm) in width and weigh 36 pounds (16 kg).

When the ray is small it grows inside its mother, positioned with wings folded over its body. It also gains nutrition from the mother's uterine secretions. It also breaks through what we call a breech birth-tail first. The cownose ray is 11 to 18 inches (280 to 460 mm) in width at birth. When it gets older it can grow to 45 inches (1.1 m) in width, and weigh 50 pounds (23 kg) or more. It is brown-backed with a whitish belly. Ovoviviparous.

Cownose rays don't have a particularly distinctive colouration but its shape is recognisable. Its eyes peer out spookily from the sides of the broad head. It also has a set of remarkable teeth plates designed for crushing clams and oyster shells.

One can also be stung by a cownose ray. The stinger is on its tail, close to the ray's body, and it doesn't usually inflict damage. The stinger is known as the spine which is pointed and it has teeth lining its lateral edges.

The cownose ray feeds upon clams. It is a voracious eater. They also eat oysters, hard clams and other invertebrates. Rays move as a group regardless of time during feeding. This group's synchronised wing flaps stir up sediment and allows the rays to find buried clams and oysters. When they locate their prey they place it in their jaws and crush it.

Cow Sharks

Cow Sharks (*Hexanchidae*) are a family of sharks characterised by extra pairs of gill slits. There are probably only two seven-gilled genera, *Heptranchias* and *Notoryhncus* (Allen, 45).

Genera and Species

- Genera *Heptranchias*
  - — Sharpnose seven-gill shark, *Heptranchias perlo* (Bonnaterre, 1788)
- Genera *Hexanchus*
  - — Bluntnose sixgill shark, *Hexanchus griseus* (Bonnaterre, 1788)
  - — Bigeye sixgill shark, *Hexanchus nakamurai* (Teng, 1962)
- Genera *Notorynchus*
  - — Broadnose sevengill shark, *Notorynchus cepedianus* (Peron, 1807)

Crestfish

Crestfishes are lampriform fishes in the family Lophotidae. They are elongate ribbon-like fishes, silver in colour, found in deep tropical and subtropical waters worldwide. Their scientific name is from Greek *lophos* meaning "crest" and refer to the crest (part of the dorsal fin) that emerges from the snout and head; this structure gives them their other name of unicorn fishes. They possess ink sacs that open into their cloaca from which they can produce a cloud of black ink when threatened (as in many cephalopods).

Species

There are three species in two genera:

- Genus *Eumecichthys*
  - — Unicorn crestfish, *Eumecichthys fiski* (Gunther, 1890).
- Genus *Lophotus*
  - — North Pacific crestfish, *Lophotus capellei* (Temminck & Schlegel, 1845).
  - — Crested oarfish, *Lophotus lacepede* (Giorna, 1809).

Sciaenidae

Sciaenidae is a family of fish commonly called drums, croakers, or hardheads for the repetitive throbbing or drumming sounds (which they make themselves) heard under water. The family includes the weakfish, and consists of about 275 species in about 70 genera; it belongs to the order Perciformes.

Sciaenids have a long dorsal fin reaching nearly to the tail, and a notch between the rays and spines of the dorsal, although the two parts are actually separate. Drums are somberly coloured, usually in shades of brown, with a lateral line that extends to the tip of the caudal fin. The anal fin usually has two spines while the dorsal fins are deeply notched or separate. Most species have a rounded or pointed caudal fin. The mouth is set low and is usually inferior.

They are found worldwide, in both fresh and saltwater, and are typically benthic carnivores, feeding on invertebrates and smaller fish. They are small to medium-sized bottom dwelling fishes that live primarily in estuaries, bays, and muddy river banks. Most of these fishes avoid clear waters such as coral reefs and oceanic islands with a few notable exceptions (i.e., Reef Croaker, High-hat, and Spotted Drum). They live in warm-temperate and tropical waters and are best represented in major rivers in S.E. Asia, N.E. South America, Gulf of Mexico, and Gulf of California.

They are excellent food and sport fishes and are commonly caught by surf and pier fishers.

The sounds are produced by the beating of abdominal muscles against the swim bladder. (In the Mississippi Valley region of the US, there is a widespread but inaccurate belief that the drumfish makes the sounds by rattling two loose bones inside its cranium. There are, in fact, no such bones.)

Genera and Selected Species

- *Aplodinotus*: Freshwater drum
- *Argyrosomus*
- *Aspericorvina*
- *Atractoscion*: White seabass
- *Atrobucca*
- *Austronibea*
- *Bahaba*
- *Bairdiella*
- *Boesemania*
- *Cheilotrema*
- *Chrysochir*
- *Cilus*
- *Collichthys*
- *Corvula*
- *Ctenosciaena*
- *Cynoscion*: Weakfish, Acoupa weakfish, Spotted seatrout
- *Daysciaena*
- *Dendrophysa*
- *Elattarchus*
- *Equetus*
- *Genyonemus*: White croaker
- *Isopisthus*
- *Johnius*
- *Kathala*
- *Larimichthys*
- *Larimus*
- *Leiostomus*: Spot croaker

- *Lonchurus*
- *Macrodon*
- *Macrospinosa*
- *Megalonibea*
- *Menticirrhus*: Kingcroaker, California corbina
- *Micropogonias*
- *Miichthys*
- *Miracorvina*
- *Nebris*
- *Nibea*
- *Odontoscion*
- *Ophioscion*
- *Otolithes*
- *Otolithoides*
- *Pachypops*
- *Pachyurus*
- *Panna*
- *Paralonchurus*
- *Paranibea*
- *Pareques*
- *Pennahia*
- *Pentheroscion*
- *Plagioscion*
- *Pogonias*
- *Protonibea*
- *Protosciaena*
- *Pseudosciaena*
- *Pseudotolithus*
- *Pteroscion*
- *Pterotolithus*

- *Roncador*: Spotfin croaker
- *Sciaena*
- *Sciaenops*: Red drum
- *Seriphus*: Queenfish
- *Sonorolux*
- *Stellifer*
- *Totoaba*
- *Umbrina*: Yellowfin croaker

Channichthyidae

The icefishes (or white-blooded fishes) are a family (Channichthyidae) of perciform fish found in the cold waters around Antarctica and southern South America.

Their blood is transparent because they have no haemoglobin and/or only defunct erythrocytes. Their metabolism relies only on the oxygen dissolved in the liquid blood, which is believed to be absorbed directly through the skin from the water. This works because water can dissolve the most oxygen when it is coldest. Also, their muscles (except the heart muscle) lack myoglobin. These extraordinary properties seem to be an adaptation to the extreme cold of their habitat. (Note that water temperature can drop below 0 °C (the freezing point of freshwater) in the Antarctic sea, but, on the other hand, stays rather constant.)

Channichthyidae are the only known vertebrates without haemoglobin. For a discussion of the discovery, genetic analysis and evolutionary implications of this condition. The chapter is titled "The Bloodless Fish of Bouvet Island". Icefish feed on krill, copepods, and other fish. There are fifteen known species of Icefish.

Crocodile Shark

The crocodile shark, *Pseudocarcharias kamoharai*, the only member of the genus *Pseudocarcharias* (in turn the only member of its family Pseudocarchariidae), is a shark found in the tropical and subtropical waters of all oceans, at depths down to 590 m. It can exceed 1 m in length.

Distribution and Habitat

The crocodile shark is an oceanic species usually found offshore and far from land but sometimes occurring inshore.

Anatomy and Appearance

It is a medium-sized spindle-shaped shark with very large eyes lacking a nictitating membrane, long gill slits extending onto the dorsal surface of the head, lanceolate teeth, weak keel and precaudal pits on the caudal peduncle. The dorsal fins are small and low, with the second dorsal fin less than half the size of the first but larger than the anal fin. The pectoral fin is broad and rounded.

The colouration is light or dark grey above, paler below, fins white-edged, sometimes with small white spots on the body, and a white blotch between the mouth and the gill slits.

Diet

Their diet is small pelagic bony fishes, squids and shrimps.

Behaviour

The behaviour of the crocodile shark is not well known, but it is believed to be a fast-swimming shark, probably capable of leaping out of the water. It is a pelagic shark. Its large eyes might indicate either nocturnal activity or feeding at great depths.

Reproduction

Reproduction is ovoviviparous, embryos feeding on yolk sac and other ova produced by the mother, with up to 4 young in a litter.

Importance to Humans

Although the crocodile shark is not considered dangerous to humans, its powerful jaws, jaw muscles and teeth invite respect. The flesh is not utilised except for its large, squalene-rich liver. It is fished by longlines off Japan but not significantly consumed.

Taxonomy

The English name is derived from the Japanese name for this animal *mizu wani,* which literally means water crocodile.

Crucian Carp

The Crucian Carp (*Carassius carassius*) is a member of the family Cyprinidae, which includes many other fish such as the common carp, or the smaller minnows. They inhabit lakes, ponds, and slow-moving rivers throughout Europe and Asia. The crucian is a medium-sized cyprinid, which rarely exceeds a weight of over 3.3 pounds (1.5 kg). They usually have a dark green back, golden sides, and reddish fins, although other colour variations exist. It is known for its ability to survive without oxygen (up to 5.5 months in winter).

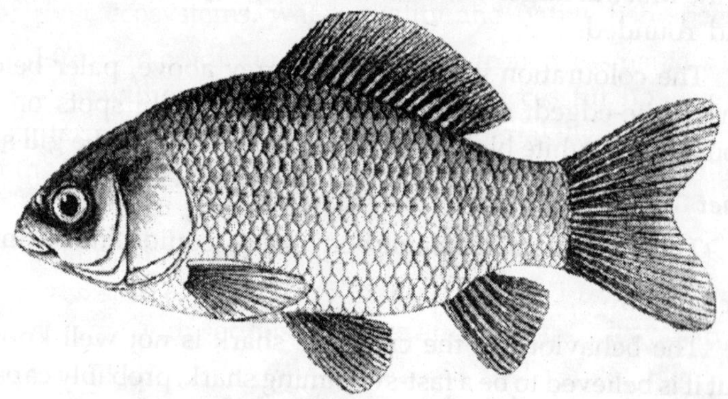

They are often caught as a sport fish the British rod-caught record for largest crucian is four pounds, nine ounces, caught by Martin Bowler in a lake in southern England in 2003. There have been various bids for a breakage of this record since, but they have been passed off as the specimens have not been said to have been "true" crucians, but hybrids between the carp and one of its relatives, such as the goldfish, which are not native to the British Isles. These hybrids often exhibit hybrid vigour or heterosis, being much more adept at finding food and evading predators than either of their parents, and thus pose somewhat of a threat to the native carp population, and to other native aquatic animals.

These carp are also occasionally kept as freshwater aquarium fish, as well as in water gardens, although they are not commonly available commercially, mainly because they are not in particularly high demand due to the presence of more colourful fish such as the koi or orfe.

The shape of a Crucian carp can be very high. The fish get an almost perfect disc shape with well rounded fins. If no predators like pike or perch are present, the Crucian carp will grow in length rather then height and the fish will be more slender looking. The growth in height will make it difficult for predators to swallow the crucian carp.

Many sources will claim that crucian carp are the wild version of the goldfish (*Carassius auratus auratus*). The wild form of the goldfish is however, *Carassius auratus gibelio*, or rather *Carassius gibelio* with *auratus* as the subspecies. While they are certainly closely related, they are different species which can be indentified by the following characteristics:

- *C. auratus* has a more pointed snout while the snout of a crucian carp is well rounded.
- The wild form of the Goldfish *C. auratus gibelio* or *C. gibelio* often has a grey/greenish colour, while crucian carps are always golden bronze.
- Juvenile crucian carp (and tench) have a black spot on the base of the tail which disappears with age. In C. auratus this tail spot is never present.
- *C. auratus* have less then 31 scales along the lateral line while crucian carp have 33 scales or more.

Ophidiidae

The Ophidiidae, including the cusk eels, are a group of marine fishes in the order Ophidiiformes. The name is from Greek *ophis* meaning "snake", and refers to their eel-like appearance. However, they can be distinguished from true eels by their ventral fins, which are developed into a forked barbel-like organ below the mouth.

They are found in temperate and tropical oceans throughout the world. The largest species, *Lamprogrammus shcherbachevi*, grows up to two metres in length, but most species are shorter than a metre.

A few species are fished commercially, most notably the pink cusk eel, *Genypterus blacodes*.

Species

There are 238 species in 49 genera:

- Subfamily Brotulinae
  - — Genus *Brotula*
    - ⇒ Bearded brotula, *Brotula barbata* (Linnaeus, 1758).
    - ⇒ Pacific bearded brotula, *Brotula clarkae* (Hubbs, 1944).
    - ⇒ Goatsbeard brotula, *Brotula multibarbata* (Temminck & Schlegel, 1846).
    - ⇒ Ordway's brotula, *Brotula ordwayi* (Hildebrand & Barton, 1949).
    - ⇒ Townsend's cusk eel, *Brotula townsendi* (Fowler, 1900).
- Subfamily Brotulotaeniinae
  - — Genus *Brotulotaenia*
    - ⇒ *Brotulotaenia brevicauda* (Cohen, 1974).
    - ⇒ Violet cuskeel, *Brotulotaenia crassa* (Parr, 1934).
    - ⇒ *Brotulotaenia nielseni* (Cohen, 1974).
    - ⇒ *Brotulotaenia nigra* (Parr, 1933).
- Subfamily Neobythitinae
  - — Genus *Abyssobrotula*
    - ⇒ *Abyssobrotula galatheae* (Nielsen, 1977).
  - — Genus *Acanthonus*
    - ⇒ Bony-eared assfish, *Acanthonus armatus* (Temminck & Schlegel, 1846).
  - — Genus *Alcockia*
    - ⇒ *Alcockia rostrata* (Gunther, 1887).

— Genus *Apagesoma*

⇒ *Apagesoma delosommatus* (Hureau, Staiger & Nielsen, 1979).

⇒ *Apagesoma edentatum* (Carter, 1983).

— Genus *Barathrites*

⇒ *Barathrites iris* (Zugmayer, 1911).

⇒ *Barathrites parri* (Nybelin, 1957).

— Genus *Barathrodemus*

⇒ *Barathrodemus manatinus* (Goode & Bean, 1883).

⇒ *Barathrodemus nasutus* (Smith & Radcliffe, 1913).

— Genus *Bassogigas*

⇒ *Bassogigas gillii* (Goode & Bean, 1896).

— Genus *Bassozetus*

⇒ Abyssal assfish, *Bassozetus compressus* (Gunther, 1878).

⇒ *Bassozetus elongatus* (Smith & Radcliffe, 1913).

⇒ *Bassozetus galatheae* (Nielsen, 1977).

⇒ *Bassozetus glutinosus* (Alcock, 1890).

⇒ *Bassozetus levistomatus* (Machida, 1989).

⇒ *Bassozetus multispinis* (Shcherbachev, 1980).

⇒ *Bassozetus nasus* (Garman, 1899).

⇒ *Bassozetus normalis* (Gill, 1883).

⇒ *Bassozetus oncerocephalus* (Vaillant, 1888).

⇒ *Bassozetus robustus* (Smith & Radcliffe, 1913).

⇒ *Bassozetus taenia* (Gunther, 1887).

⇒ *Bassozetus werneri* (Nielsen & Merrett, 2000).

⇒ *Bassozetus zenkevitchi* (Rass, 1955).

— Genus *Bathyonus*

⇒ *Bathyonus caudalis* (Garman, 1899).

⇒ *Bathyonus laticeps* (Gunther, 1878).

⇒ *Bathyonus pectoralis* (Goode & Bean, 1885).

— Genus *Benthocometes*

⇒ *Benthocometes robustus* (Smith & Radcliffe, 1913).

— Genus *Dannevigia*

⇒ Australian tusk, *Dannevigia tusca* (Whitley, 1941).

— Genus *Dicrolene*

⇒ *Dicrolene filamentosa* (Garman, 1899).

⇒ *Dicrolene gregoryi* (Trotter, 1926).

⇒ *Dicrolene hubrechti* (Weber, 1913).

⇒ Digitate cusk eel, *Dicrolene introniger* (Goode &Bean, 1883).

⇒ *Dicrolene kanazawai* (Grey, 1958).

⇒ *Dicrolene longimana* (Smith & Radcliffe, 1913).

⇒ *Dicrolene mesogramma* (Shcherbachev, 1980).

⇒ Slender brotula, *Dicrolene multifilis* (Alcock, 1889).

⇒ *Dicrolene nigra* (Parr, 1933).

⇒ *Dicrolene nigricaudis* (Alcock, 1891).

⇒ *Dicrolene pallidus* (Hureau & Nielsen, 1981).

⇒ *Dicrolene pullata* (Garman, 1899.

⇒ *Dicrolene quinquarius* (Gunther, 1887).

⇒ *Dicrolene tristis* (Smith & Radcliffe, 1913).

⇒ *Dicrolene vaillanti* (Alcock, 1890).

— Genus *Enchelybrotula*

⇒ *Enchelybrotula gomoni* (Cohen, 1982).

⇒ *Enchelybrotula paucidens* (Smith & Radcliffe, 1913).

— Genus *Epetriodus*

⇒ *Epetriodus freddyi* (Cohen & Nielsen, 1978).

— Genus *Eretmichthys*

⇒ *Eretmichthys pinnatus* (Garman, 1899).

— Genus *Glyptophidium*

⇒ *Glyptophidium argenteum* (Alcock, 1889).

⇒ *Glyptophidium effulgens* (Nielsen & Machida, 1988).

⇒ *Glyptophidium japonicum* (Steindachner & Doderlein, 1887).

⇒ Bigeye brotula, *Glyptophidium longipes* (Norman, 1939).

⇒ *Glyptophidium lucidum* (Smith & Radcliffe, 1913).

⇒ *Glyptophidium macropus* (Alcock, 1894).

⇒ *Glyptophidium oceanium* (Smith & Radcliffe, 1913).

— Genus *Holcomycteronus*

⇒ *Holcomycteronus aequatoris* (Smith & Radcliffe, 1913).

⇒ *Holcomycteronus brucei* (Dollo, 1906).

⇒ *Holcomycteronus digittatus* (Garman, 1899).

⇒ *Holcomycteronus profundissimus* (Roule, 1913).

⇒ *Holcomycteronus pterotus* (Alcock, 1890).

⇒ *Holcomycteronus squamosus* (Roule, 1916).

— Genus *Homostolus*

⇒ *Homostolus acer* (Smith & Radcliffe, 1913).

— Genus *Hoplobrotula*

⇒ *Hoplobrotula armata* (Temminck & Schlegel, 1846).

⇒ *Hoplobrotula badia* (Machida, 1990).

⇒ False kinglip, *Hoplobrotula gnathopus* (Regan, 1921).

— Genus *Hypopleuron*

⇒ *Hypopleuron caninum* (Smith & Radcliffe, 1913).

— Genus *Lamprogrammus*

⇒ *Lamprogrammus brunswigi* (Brauer, 1906).

⇒ Legless cuskeel, *Lamprogrammus exutus* (Nybelin & Poll, 1958).

⇒ *Lamprogrammus fragilis* (Alcock, 1892).

⇒ *Lamprogrammus niger* (Alcock, 1891).

⇒ *Lamprogrammus shcherbachevi* (Cohen & Rohr, 1993).

— Genus *Leptobrotula*

⇒ *Leptobrotula breviventralis* (Nielsen, 1986).

— Genus *Leucicorus*
   ⇒ *Leucicorus atlanticus* (Nielsen, 1975).
   ⇒ *Leucicorus lusciosus* (Garman, 1899).

— Genus *Luciobrotula*
   ⇒ *Luciobrotula bartschi* (Smith & Radcliffe, 1913).
   ⇒ *Luciobrotula corethromycter* (Cohen, 1964).
   ⇒ *Luciobrotula lineata* (Gosline, 1954).
   ⇒ *Luciobrotula nolfi* (Cohen, 1981.

— Genus *Mastigopterus*
   ⇒ *Mastigopterus imperator* (Smith & Radcliffe, 1913).

— Genus *Monomitopus*
   ⇒ *Monomitopus agassizii* (Goode & Bean, 1896).
   ⇒ *Monomitopus americanus* (Nielsen, 1971).
   ⇒ *Monomitopus conjugator* (Alcock, 1896).
   ⇒ *Monomitopus garmani* (Smith & Radcliffe, 1913).
   ⇒ *Monomitopus kumae* (Jordan & Hubbs, 1925).
   ⇒ *Monomitopus longiceps* (Smith & Radcliffe, 1913).
   ⇒ *Monomitopus magnus* (Carter & Cohen, 1985).
   ⇒ *Monomitopus malispinosus* (Garman, 1899).
   ⇒ *Monomitopus metriostoma* (Vaillant, 1888).
   ⇒ *Monomitopus microlepis* (Matsubara, 1943).
   ⇒ *Monomitopus nigripinnis* (Alcock, 1889).
   ⇒ *Monomitopus pallidus* (Hureau & Nielsen, 1981).
   ⇒ *Monomitopus torvus* (Garman, 1899).
   ⇒ *Monomitopus vitiazi* (Nielsen, 1971).

— Genus *Neobythites*
   ⇒ *Neobythites alcocki* (Nielsen, 2002).
   ⇒ Black-edged cuskeel, *Neobythites analis* (Barnard, 1927).
   ⇒ *Neobythites andamanensis* (Nielsen, 2002).
   ⇒ *Neobythites australiensis* (Nielsen, 2002).
   ⇒ *Neobythites bimaculatus* (Nielsen, 1997).

⇒ *Neobythites bimarginatus* (Fourmanoir & Rivaton, 1979).
⇒ *Neobythites braziliensis* (Nielsen, 1999).
⇒ *Neobythites crosnieri* (Nielsen, 1995).
⇒ *Neobythites fasciatus* (Smith & Radcliffe, 1913).
⇒ *Neobythites fijiensis* (Nielsen, 2002).
⇒ *Neobythites franzi* (Nielsen, 2002).
⇒ Twospot brotula, *Neobythites gilli* (Goode & Bean, 1896).
⇒ *Neobythites javaensis* (Nielsen, 2002).
⇒ *Neobythites kenyaensis* (Nielsen, 1995).
⇒ *Neobythites longipes* (Norman, 1939).
⇒ *Neobythites longispinis* (Nielsen, 2002).
⇒ *Neobythites longiventralis* (Nielsen, 1997).
⇒ *Neobythites macrocelli* (Nielsen, 2002).
⇒ *Neobythites macrops* (Gunther, 1887).
⇒ *Neobythites malayanus* (Weber, 1913).
⇒ *Neobythites malhaensis* (Nielsen, 1995).
⇒ Stripefin brotula, *Neobythites marginatus* (DeKay, 1842).
⇒ *Neobythites marianaensis* (Nielsen, 2002).
⇒ *Neobythites marquesaensis* (Nielsen, 2002).
⇒ *Neobythites meteori* (Nielsen, 1995).
⇒ *Neobythites monocellatus* (Nielsen, 1999).
⇒ *Neobythites multidigitatus* (Nielsen, 1999).
⇒ *Neobythites multistriatus* (Nielsen & Quero, 1991).
⇒ *Neobythites musorstomi* (Nielsen, 2002).
⇒ *Neobythites natalensis* (Nielsen, 1995).
⇒ *Neobythites neocaledoniensis* (Nielsen, 1997).
⇒ *Neobythites nigriventris* (Nielsen, 2002).
⇒ *Neobythites ocellatus* (Garman, 1899).
⇒ *Neobythites pallidus* (Hureau & Nielsen, 1981).

⇒ *Neobythites purus* (Smith & Radcliffe, 1913).

⇒ *Neobythites sereti* (Nielsen, 2002).

⇒ *Neobythites sinensis* (Nielsen, 2002).

⇒ *Neobythites sivicola* (Jordan & Snyder, 1901).

⇒ *Neobythites soelae* (Nielsen, 2002).

⇒ *Neobythites somaliaensis* (Nielsen, 1995).

⇒ *Neobythites steatiticus* (Alcock, 1894).

⇒ *Neobythites stefanovi* (Nielsen & Uiblein, 1993).

⇒ Thread brotula, *Neobythites stelliferoides* (Gilbert, 1890).

⇒ *Neobythites stigmosus* (Machida, 1984).

⇒ *Neobythites trifilis* (Kotthaus, 1979).

⇒ *Neobythites unicolor* (Nielsen & Retzer, 1994).

⇒ *Neobythites unimaculatus* (Smith & Radcliffe, 1913).

⇒ *Neobythites vityazi* (Nielsen, 1995).

⇒ *Neobythites zonatus* (Nielsen, 1997).

— Genus *Neobythitoides*

⇒ *Neobythitoides serratus* (Nielsen & Machida, 2006).

— Genus *Penopus*

⇒ *Penopus microphthalmus* (Vaillant, 1888).

— Genus *Petrotyx*

⇒ Velvetnose brotula, *Petrotyx hopkinsi* (Heller & Snodgrass, 1903).

⇒ Redfin brotula, *Petrotyx sanguineus* (Meek & Hildebrand, 1928).

— Genus *Porogadus*

⇒ *Porogadus abyssalis* (Nybelin, 1957).

⇒ *Porogadus atripectus* (Garman, 1899).

⇒ *Porogadus catena* (Goode & Bean, 1885).

⇒ Cavernous assfish, *Porogadus gracilis* (Gunther, 1878).

⇒ *Porogadus guentheri* (Jordan & Fowler, 1902).

⇒ *Porogadus longiceps* (Smith & Radcliffe, 1913).

⇒ *Porogadus melampeplus* (Alcock, 1896).

⇒ *Porogadus melanocephalus* (Alcock, 1891).

⇒ Slender cuskeel, *Porogadus miles* (Goode & Bean, 1885).

⇒ *Porogadus nudus* (Vaillant, 1888).

⇒ *Porogadus silus* (Carter & Sulak, 1984).

⇒ *Porogadus subarmatus* (Vaillant, 1888).

⇒ *Porogadus trichiurus* (Alcock, 1890).

— Genus *Pycnocraspedum*

⇒ *Pycnocraspedum armatum* (Temminck & Schlegel, 1846).

⇒ *Pycnocraspedum fulvum* (Hildebrand & Barton, 1949).

⇒ *Pycnocraspedum microlepis* (Matsubara, 1943).

⇒ *Pycnocraspedum phyllosoma* (Parr, 1933).

⇒ *Pycnocraspedum squamipinne* (Alcock, 1889).

— Genus *Selachophidium*

⇒ Gunther's cuskeel, *Selachophidium guentheri* (Jordan & Fowler, 1902).

— Genus *Sirembo*

⇒ *Sirembo imberbis* (Temminck & Schlegel, 1846).

⇒ Brown-banded cusk eel, *Sirembo jerdoni* (Day, 1888).

⇒ *Sirembo marmoratum* (Goode & Bean, 1885).

⇒ *Sirembo metachroma* (Cohen & Robins, 1986).

— Genus *Spectrunculus*

⇒ Pudgy cuskeel, *Spectrunculus grandis* (Gunther, 1877).

— Genus *Spottobrotula*

⇒ *Spottobrotula amaculata* (Cohen & Nielsen, 1982).

⇒ *Spottobrotula mahodadi* (Cohen & Nielsen, 1978).

— Genus *Tauredophidium*

⇒ *Tauredophidium hextii* (Alcock, 1890).

— Genus *Typhlonus*

⇒ *Typhlonus nasus* (Garman, 1899).

— Genus *Ventichthys*

⇒ *Ventichthys biospeedoi* (Nielsen, Møller & Segonzac, 2006).

— Genus *Xyelacyba*

⇒ *Xyelacyba myersi* (Cohen, 1961).

- Subfamily Ophidiinae

— Genus *Cherublemma*

⇒ Black brotula, *Cherublemma emmelas* (Gilbert, 1890).

— Genus *Chilara*

⇒ Spotted cusk eel, *Chilara taylori* (Girard, 1858).

— Genus *Genypterus*

⇒ Pink cusk eel, *Genypterus blacodes* (Forster, 1801).

⇒ *Genypterus brasiliensis* (Regan, 1903).

⇒ Kingklip, *Genypterus capensis* (Smith, 1847).

⇒ Red cusk eel, *Genypterus chilensis* (Guichenot, 1848).

⇒ Black cusk eel, *Genypterus maculatus* (Kaup, 1858).

⇒ Rock ling, *Genypterus tigerinus* (Klunzinger, 1872).

— Genus *Lepophidium*

⇒ Dusky cusk eel, *Lepophidium aporrhox* (Robins, 1961).

⇒ Shortbeard cusk eel, *Lepophidium brevibarbe* (Cuvier, 1829).

⇒ Mottled cusk eel, *Lepophidium jeannae* (Fowler, 1941).

⇒ *Lepophidium kallion* (Robins, 1959).

⇒ Silver cusk eel, *Lepophidium microlepis* (Matsubara, 1943).

- ⇒ Specklefin cusk eel, *Lepophidium negropinna* (Hildebrand & Barton, 1949).
- ⇒ Leopard cusk eel, *Lepophidium pardale* (Gilbert, 1890).
- ⇒ Blackedge cusk eel, *Lepophidium pheromystax* (Robins, 1960).
- ⇒ Blackrim cusk eel, *Lepophidium profundorum* (Gill, 1863).
- ⇒ Prowspine cusk eel, *Lepophidium prorates* (Jordan & Bollman, 1890).
- ⇒ Barred cusk eel, *Lepophidium staurophor* (Robins, 1958).
- ⇒ Blotchfin cusk eel, *Lepophidium stigmatistium* (Gilbert, 1890).

— Genus *Ophidion*

- ⇒ Longnose cusk eel, *Ophidion antipholus* (Lea & Robins, 2003).
- ⇒ *Ophidion asiro* (Jordan & Fowler, 1902).
- ⇒ Snake blenny, *Ophidion barbatum* (Linnaeus, 1758).
- ⇒ *Ophidion beani* (Jordan & Gilbert, 1883).
- ⇒ Shorthead cusk eel, *Ophidion dromio* (Lea & Robins, 2003).
- ⇒ *Ophidion exul* (Robins, 1991).
- ⇒ Earspot cusk eel, *Ophidion fulvum* (Hildebrand & Barton, 1949).
- ⇒ Striped cusk eel, *Ophidion galeoides* (Gilbert, 1890).
- ⇒ *Ophidion genyopus* (Ogilby, 1897).
- ⇒ Blotched cusk eel, *Ophidion grayi* (Fowler, 1948).
- ⇒ Band cusk eel, *Ophidion holbrookii* (Putnam, 1874).
- ⇒ Mimic cusk eel, *Ophidion imitator* (Lea, 1997).
- ⇒ Rainbow cusk eel, *Ophidion iris* (Zugmayer, 1911).
- ⇒ *Ophidion josephi* (Girard, 1858).
- ⇒ Harelip cusk, *Ophidion lagochila* (Bohlke & Robins, 1959).

⇒ *Ophidion lozanoi* (Matallanas, 1990).

⇒ Striped cusk eel, *Ophidion marginatum* (DeKay, 1842).

⇒ *Ophidion metoecus* (Robins, 1991).

⇒ *Ophidion muraenolepis* (Gunther, 1880).

⇒ Letter opener, *Ophidion nocomis* (Robins & Bohlke, 1959).

⇒ Colonial cusk eel, *Ophidion robinsi* (Fahay, 1992).

⇒ *Ophidion rochei* (Muller, 1845).

⇒ *Ophidion saldanhai* (Matallanas & Brito, 1999).

⇒ Basketweave cusk eel, *Ophidion scrippsae* (Hubbs, 1916).

⇒ Mooneye cusk eel, *Ophidion selenops* (Robins & Bohlke, 1959).

⇒ *Ophidion smithi* (Fowler, 1934).

⇒ Crested cusk eel, *Ophidion welshi* (Nichols & Breder, 1922).

— Genus *Otophidium*

⇒ Ghost cusk eel, *Otophidium chickcharney* (Bohlke & Robins, 1959).

⇒ Sleeper cusk eel, *Otophidium dormitator* (Bohlke & Robins, 1959).

⇒ Bighead cusk eel, *Otophidium indefatigabile* (Jordan & Bollman, 1890).

⇒ Polka-dot cusk eel, *Otophidium omostigma* (Jordan & Gilbert, 1882).

— Genus *Parophidion*

⇒ Dusky cusk eel, *Parophidion schmidti* (Woods & Kanazawa, 1951).

⇒ *Parophidion vassali* (Risso, 1810).

— Genus *Raneya*

⇒ *Raneya brasiliensis* (Regan, 1903).

Cutlassfish

The cutlassfishes are about 40 species of fish in the family Trichiuridae (order Perciformes) found in seas throughout the world. Fish of this family are long, slender, and generally steely blue or silver in colour, giving rise to their name. Some species are known as scabbardfishes or hairtails or Walla Walla; others are called frostfishes because they appear in late autumn and early winter, around the time of the first frosts.

Classification

This list of species follows FishBase:

- Genus *Aphanopus*
  - — *Aphanopus arigato* (Parin, 1994)
  - — *Aphanopus beckeri* (Parin, 1994)
  - — *Aphanopus capricornis* (Parin, 1994)
  - — Black scabbardfish, *Aphanopus carbo* (Lowe, 1839)
  - — Intermediate scabbardfish, *Aphanopus intermedius* (Parin, 1983)
  - — Smalleye scabbardfish, *Aphanopus microphthalmus* (Norman, 1939)
  - — Mikhailin's scabbardfish, *Aphanopus mikhailini* (Parin, 1983)
- Genus *Assurger*
  - — Razorback scabbardfish, *Assurger anzac* (Alexander, 1917)
- Genus *Benthodesmus*
  - — Elongate frostfish, *Benthodesmus elongatus* (Clarke, 1879)
  - — Bigeye frostfish, *Benthodesmus macrophthalmus* (Parin & Becker, 1970)
  - — Neglected frostfish, *Benthodesmus neglectus* (Parin, 1976)
  - — Sparse-rayed frostfish, *Benthodesmus oligoradiatus* (Parin & Becker, 1970)

— North-Pacific frostfish, *Benthodesmus pacificus* (Parin & Becker, 1970)
— Papuan frostfish, *Benthodesmus papua* (Parin, 1978)
— Simony's frostfish, *Benthodesmus simonyi* (Steindachner, 1891)
— Philippine frostfish, *Benthodesmus suluensis* (Parin, 1976)
— Slender frostfish, *Benthodesmus tenuis* (Gunther, 1877)
— Tucker's frostfish, *Benthodesmus tuckeri* (Parin & Becker, 1970)
— Vityaz' frostfish, *Benthodesmus vityazi* (Parin & Becker, 1970)

- Genus *Demissolinea*

— New Guinean hairtail, *Demissolinea novaeguineensis* (Burhanuddin & Iwatsuki, 2003)

- Genus *Eupleurogrammus*

— Longtooth hairtail, *Eupleurogrammus glossodon* (Bleeker, 1860)
— Smallhead hairtail, *Eupleurogrammus muticus* (Gray, 1831)

- Genus *Evoxymetopon*

— *Evoxymetopon macrophthalmus* (Chakraborty, Yoshino & Iwatsuki, 2006).
— Poey's scabbardfish, *Evoxymetopon poeyi* (Gunther, 1887)
— Channel scabbardfish, *Evoxymetopon taeniatus* (Gill, 1863)

- Genus *Lepidopus*

— Crested scabbardfish, *Lepidopus altifrons* (Parin & Collette, 1993)
— Hawaiian ridge scabbardfish, *Lepidopus calcar* (Parin & Mikhailin, 1982)
— Silver scabbardfish, *Lepidopus caudatus* (Euphrasen, 1788)
— Doubtful scabbardfish, *Lepidopus dubius* (Parin & Mikhailin, 1981)

- Fitch's scabbardfish, *Lepidopus fitchi* (Rosenblatt & Wilson, 1987)
- Ghost scabbardfish, *Lepidopus manis* (Rosenblatt & Wilson, 1987)

• Genus *Lepturacanthus*

- Coromandel hairtail, *Lepturacanthus pantului* (Gupta, 1966)
- Savalani hairtail, *Lepturacanthus savala* (Cuvier, 1829)

• Genus *Tentoriceps*

- Crested hairtail, *Tentoriceps cristatus* (Klunzinger, 1884)

• Genus *Trichiurus*

- Pearly hairtail, *Trichiurus auriga* (Klunzinger, 1884)
- *Trichiurus australis* (Chakraborty, Burhanuddin & Iwatsuki, 2005).
- Chinese short-tailed hairtail, *Trichiurus brevis* (Wang & You, 1992)
- Ganges hairtail, *Trichiurus gangeticus* (Gupta, 1966)
- Largehead hairtail, *Trichiurus lepturus* (Linnaeus, 1758)
- *Trichiurus margarites* (Li, 1992)
- *Trichiurus nanhaiensis* (Wang & Xu, 1992)
- Australian short tailed hairtail, *Trichiurus nickolensis* (Burhanuddin & Iwatsuki, 2003)
- Short-tailed hairtail, *Trichiurus russelli* (Dutt & Thankam, 1966)

Cutthroat eel

Cutthroat eels are a family, Synaphobranchidae, of eels, the only member of the suborder Synaphobranchoidei. They are native to the Atlantic, Pacific and Indian oceans.

Genera and Species

There are 36 species in 12 genera:

• Genus *Atractodenchelys*

• Genus *Diastobranchus*

- Basketwork eel, *Diastobranchus capensis*

- Genus *Dysomma*
  - — *Dysomma anguillare*
  - — *Dysomma longirostrum*
  - — *Dysomma melanurum*
- Genus *Dysommina*
- Genus *Haptenchelys*
- Genus *Histiobranchus*
  - — Deep-water arrowtooth eel, *Histiobranchus bathybius*
  - — Bruun's cutthroat eel, *Histiobranchus bruuni*
- Genus *Ilyophis*
  - — Muddy arrowtooth eel, *Ilyophis brunneus*
- Genus *Linkenchelys*
- Genus *Meadia*
  - — *Meadia abyssalis*
  - — *Meadia roseni*
- Genus *Simenchelys*
  - — Snubnosed eel, *Simenchelys parasitica*
- Genus *Synaphobranchus*
  - — Grey cutthroat eel, *Synaphobranchus. affinis*
- Genus *Thermobiotes*

In some classifications (for example, ITIS), this family is split, with *Simenchelys* in its own family, Simenchelyidae.

Cutthroat Trout

The cutthroat trout (*Oncorhynchus clarki*) is a species of freshwater fish in the salmon family (family Salmonidae) of order Salmoniformes. It is one of the many trout.

Cutthroat trout are native to western North America. Some populations live primarily in the Pacific Ocean as adults and return to fresh water from fall through early spring, to feed on freshwater insects and to spawn, however most populations stay in freshwater throughout their lives. At least three subspecies are confined to isolated basins in the Great Basin and can tolerate saline or alkaline water. All subspecies of cutthroat trout are sought after gamefish, especially via fly fishing.

Cutthroat are similar to rainbow trout and will readily interbreed, producing fertile hybrids. Cutthroat trout vary widely in size, colouration, and habitats. Though their colouration can range from golden to gray to green on their backs, depending on subspecies and habitat, all populations feature red, pink, or orange marks on the bottom of their jaws, which are the easiest diagnostic of the species for the casual observer. As adults, different populations and subspecies of cutthroat can range from six to almost forty inches in length, which means size is not an effective indicator of the species.

There are about 14 separate subspecies of cutthroat trout, including:

- Alvord cutthroat trout, *Oncorhynchus clarki alvordensis*; possibly extinct) Named in 2002.
- Bonneville cutthroat trout, *Oncorhynchus clarki utah*
- Coastal cutthroat trout, *Oncorhynchus clarki clarki*
- Colorado River cutthroat trout, *Oncorhynchus clarki pleuriticus*
- Crescenti trout, *Salmo clarkii crescentis*
- Greenback cutthroat trout, *Oncorhynchus clarki stomias* (threatened)
- Lahontan cutthroat trout, *Oncorhynchus clarki henshawi* (threatened)
- Paiute cutthroat trout, *Oncorhynchus clarki seleniris* (threatened)
- Rio Grande cutthroat trout, *Oncorhynchus clarki virginalis*
- Snake River fine-spotted cutthroat trout, *Oncorhynchus clarki* subsp. Named *Oncorhynchus clarki behnkei*, 1995 & 2002.
- Westslope cutthroat trout, *Oncorhynchus clarki lewisi*
- Yellowfin cutthroat trout, *Oncorhynchus clarki macdonaldi* (extinct)
- Yellowstone cutthroat trout, *Oncorhynchus clarki bouvieri*

Common Dab

The common dab is an edible flatfish in the genus Limanda. Like other flatfishes, it has both eyes on the same side of the head. It is usually under 30 cm long and under 1 kg in weight. It lives on sandy seabeds. It is most common in the North Sea.

Dace

A dace is any of a number of species of small cyprinid fish. The unmodified name is usually a reference to the common dace (*Leuciscus leuciscus*) but it may also mean:

- Fallfish, *Semotilus corporalis* (Canada)
- Chinese mud carp *Cirrhinus chinensis* (Hong Kong)

"Dace" also appears in combination to designate a variety of species:

- Blacknose dace, *Rhinichthys atratulus*
- Blackside dace, *Phoxinus cumberlandensis*
- Columbia River dace, *Ptychocheilus oregonensis*
- Danilevskii's dace, *Leuciscus danilewskii*
- Desert dace, *Eremichthys acros*
- Eurasian dace, *Leuciscus leuciscus*
- Finescale dace, *Phoxinus neogaeus*
- Horned dace, *Semotilus atromaculatus*
- Korean splendid dace, *Coreoleuciscus splendidus*
- Issyk-Kul dace, *Leuciscus bergi*
- Lake Candidus dace, *Candidia barbatus*
- Las Vegas dace, *Rhinichthys deaconi*
- Leopard dace, *Rhinichthys falcatus*
- Longfin dace, *Rhinichthys chrysogaster*
- Longnose dace, *Rhinichthys cataractae*
- Mexican dace, *Evarra bustamantei / eigenmanni / tlahuacensis*
- Moapa dace, *Moapa coriacea*
- Mountain blackside dace, *Phoxinus cumberlandensis*

- Mountain redbelly dace, *Phoxinus oreas*
- Nachtrieb dace, *Margariscus margarita*
- Northern dace, *Margariscus margarita*
- Northern pearl dace, *Margariscus margarita*
- Northern redbelly dace, *Phoxinus eos*
- Northwest dace, *Mylocheilus caurinus*
- Pearl dace, *Margariscus margarita*
- Redbelly dace, *Phoxinus eos*
- Relict dace, *Relictus solitarius*
- Rosyside dace, *Clinostomus funduloides*
- Saskatchewan dace, *Platygobio gracilis*
- Schmidt's dace, *Leuciscus schmidti*
- Shining dace, *Semotilus corporalis*
- Siberian dace, *Leuciscus baicalensis*
- Southern redbelly dace, *Phoxinus erythrogaster*
- Speckled dace, *Rhinichthys osculus*
- Syr-Darya dace, *Leuciscus squaliusculus*
- Tennessee dace, *Phoxinus tennesseensis*
- Umatilla dace, *Rhinichthys umatilla*
- Umpqua dace, *Rhinichthys evermanni*
- Zeravshan dace, *Leuciscus lehmanni*

Daggertooth

The North Pacific daggertooth, *Anotopterus nikparini,* is a daggertooth, the only member of the genus *Anotopterus,* which is the only genus in the family Anotopteridae. It is found in the North Pacific, above latitude 25°N, to depths up to and exceeding 2,000 m.

The North Pacific daggertooth is an extremely elongate scaleless fish. Large adults have a pair of dermal keels midlaterally on each side. There is a small, flexible projection at the tip of the lower jaw, and there is no rayed dorsal fin.

North Pacific daggertooth are found in a wide range of depths, sometimes near the surface to below 2,000 m. Larger adults inhabit colder water towards the poles, whereas the young and smaller adults inhabit more temperate regions. They feed on molluscs, crustaceans, marine worms, coelenterates, salps, and fishes. The distensible body wall and large stomach can accommodate prey up to half of its own length. Reproduction is oviparous, with planktonic larvae. They are preyed upon by albacore, Pacific lancetfish, halibut, steelhead salmon, blue shark, pomfret, and whales.

Daggertooth Pike Conger

The daggertooth pike conger or pike eel, *Muraenesox cinereus,* is a type of eel. It is a traditional food in Japanese cuisine. It has a lot of bones.

Damselfish

Damselfish refers to members of the family Pomacentridae, except those of the two genera Amphiprion and Premnasmost, most usually *Chromis chromis*. Other species within the family have common names that include the word 'damselfish', but in almost all cases this is qualified with an adjective or other descriptor.

The average size of such damselfish is around 3 inches (8 centimetres). They are all marine, however, a couple of species are regularly found in the lower stretches of rivers in pure freshwater, and usually have bright colours. Some species of damselfish are able to adapt well in an average aquarium, but others such as the white-spotted damselfish cannot. The diet of a damselfish can include small crustaceans, plankton, and algae.

Many species of damselfish are kept as aquaria, and live in tropical coral reefs; however, many also live in temperate climates. One example would be damsels inhabiting the coast of southern California and northern pacific Mexican coast.

A common function for Damselfish is as a biological stabiliser in new aquariums. The fish would live in the aquarium during its initial existence, and be used to allow the aquarium to biologically stabilise with beneficial bacteria. This practice is viewed negatively by many aquarists because of the foul conditions the fish are subjected to and the fact that more humane methods to stabilise an aquarium exist.

Danio

The *danio* genus comprises many of the species of danionins familiar to aquarists. The common name "danio" is used for members of the genera *danio* as well as *Devario.*

Characteristics

They are native to the fresh water rivers and streams of South East Asia, but many species are brightly coloured, and are available as aquarium fish worldwide. A number of the species have only been recently discovered, in remote inland areas of Myanmar and do not yet have scientific names.

They have two pairs of long barbels and are generally characterised by horizontal stripes (with the exception of the Glowlight danio, Panther danio and Black Barred danio which have vertical bars). They range from 4 cm/ 1.75 in) to 15 cm/ 6 in) in length. They generally do not live for more than two to three years and are probably annual fish in the wild.

In the wild, these fish consume various small aquatic insects, crustaceans, worms, as well as, in the case of fry plankton.

Aquarium Care

The care of members of the genus *danio* are rather similar and easily generalised. They are easy to moderate in difficulty to keep.

All of these fish are primarily surface feeders. They are omnivorous in the aquarium and will accept a wide variety of foods, though flake food is appropriate. Living in aquaria, live/frozen flaked foods are suitable, especially brine shrimp and sinking tablets. When conditioning danios for breeding, it's advisable to feed them plenty of fresh foods.

Although boisterous and liable to chase each other and

other fish, they are good community fish and will not generally attack each other or other fish, although they occasionally nip fins, more by accident than design and will, like most fish, eat eggs and any fish small enough to fit into their mouths.

They are best kept in a tank long enough for their active swimming, preferably with a current from a power filter (or at least airstone) as they often live in fast flowing streams in the wild. Generally this also results in them being sub tropical with cooler temperatures. They are good jumpers and a tight fitting lid is recommended

Schooling fish, they prefer to be in groups of six or more. Danios prefer water with a 6.0-8.0 pH, a water hardness of up to 5.0-19.0 dGH,a carbon hardness of 8 to 12 KH, and a temperature range of 68-80 °F (18-24 °C), the lower end of the temperature range is ideal.

Breeding

Some species of danios, such as the Zebra danio are among the easiest aquarium fish to breed. Other species such as danio kyathit are far harder to spawn. All scatter their eggs over the substrate. The eggs are non- adhesive and hatch within 2-3 days. Eggs will be eaten enthusiastically unless protected by a layer of marbles or heavy substrate planting.

Clades

There are four clades or subgroups within the danio genus:

- *Danio dangila clade*
  - — Danio dangila
  - — Danio feegradei
  - — Danio meghalayensis
  - — Danio sp "TW01"
  - — Danio sp "KP01"
- *Danio rerio clade*
  - — Danio rerio
  - — Danio rerio var frankei

  - Danio nigrofasciatus
  - Danio kyathit var striped
  - Danio kyathit var spotted
  - Danio sp aff kyathit (sp "redfin")
  - Danio sp "TW02"
- *Danio albolieatus clade*
  - Danio albolineatus
  - Danio abolineatus var pulcher
  - Danio abolineatus var tweediei
  - Danio kerri
  - Danio roseus
  - Danio sp "Hikari"
- *Danio choprae clade*
  - Danio choprae
  - Danio choprae var putao
  - Danio sp "TW03"

It is possible that the danio dangila clade may be separated into a separate genus from the others, owing to their larger size, etc., however other than larger size and larger egg size they are very similar to the other danios in anatomy and behaviour so this is unlikely at the present time. If they were split off they would retain the title "danio" with the other danios renamed, because the "type" of danio (first discovered danio) is danio dangila, which is within this clade.

There has been speciation as to whether danio choprae was a true danio or a devario, however the discovery of danio sp "TW03" and danio choprae var putao has made this less likely.

Danios and Devarios

Most species that were formerly within the *danio* genus such as the Giant danio have now been reclassified into the *Devario* genus. For full details of the classification of *danios* and *Devarios* are on the danionin page.

Etheostoma

The Etheostoma are a genus of small freshwater fish. Most are native to North America. They are commonly known as darters. Many can produce Schreckstoff substances that serve to warn nearby fish in case of an attack.

Species List

- Alabama darter, *Etheostoma ramseyi*
- Arkansas darter, *Etheostoma cragini*
- Arkansas saddled darter, *Etheostoma euzonum*
- Arrow darter, *Etheostoma sagitta*
- Ashy darter, *Etheostoma cinereum*
- Backwater darter, *Etheostoma zonifer*
- Banded darter, *Etheostoma zonale*
- Bandfin darter, *Etheostoma zonistium*
- Barcheek darter, *Etheostoma obeyense*
- Barrens darter, *Etheostoma forbesi*
- Bayou darter, *Etheostoma rubrum*
- Blackside snubnose darter, *Etheostoma duryi*
- Blackfin darter, *Etheostoma nigripinne*
- Blenny darter, *Etheostoma blennius*
- Bloodfin darter, *Etheostoma sanguifluum*
- Bluebreast darter, *Etheostoma camurum*
- Blueside darter, *Etheostoma jessiae*
- Bluntnose darter, *Etheostoma chlorosoma*
- Boulder darter, *Etheostoma wapiti*
- Brighteye darter, *Etheostoma lynceum*
- Brook darter, *Etheostoma burri*
- Brown darter, *Etheostoma edwini*
- Buffalo darter, *Etheostoma bison*
- Candy darter, *Etheostoma osburni*

- Carolina darter, *Etheostoma collis*
- Cherokee darter, *Etheostoma scotti*
- Cherry darter, *Etheostoma etnieri*
- Chickasaw darter, *Etheostoma cervus*
- Choctawhatchee darter, *Etheostoma davisoni*
- Christmas darter, *Etheostoma hopkinsi*
- Coastal darter, *Etheostoma colorosum*
- Coldwater darter, *Etheostoma ditrema*
- Conchos darter, *Etheostoma australe*
- Coosa darter, *Etheostoma coosae*
- Coppercheek darter, *Etheostoma aquali*
- Corrugated darter, *Etheostoma basilare*
- Creole darter, *Etheostoma collettei*
- Crown darter, *Etheostoma corona*
- Cumberland darter, *Etheostoma susanae*
- Current darter, *Etheostoma uniporum*
- Cypress darter, *Etheostoma proeliare*
- Duskytail darter, *Etheostoma percnurum*
- Egg-mimic darter, *Etheostoma pseudovulatum*
- Emerald darter, *Etheostoma baileyi*
- Etowah darter, *Etheostoma etowahae*
- Fantail darter, *Etheostoma flabellare*
- Firebelly darter, *Etheostoma pyrrhogaster*
- Fountain darter, *Etheostoma fonticola*
- Fringed darter, *Etheostoma crossopterum*
- Glassy darter, *Etheostoma vitreum*
- Golden darter, *Etheostoma denoncourti*
- Goldstripe darter, *Etheostoma parvipinne*
- Greenbreast darter, *Etheostoma jordani*
- Greenfin darter, *Etheostoma chlorobranchium*

- Greenside darter, *Etheostoma blennioides*
- Greenthroat darter, *Etheostoma lepidum*
- Guardian darter, *Etheostoma oophylax*
- Gulf darter, *Etheostoma swaini*
- Harlequin darter, *Etheostoma histrio*
- Headwater darter, *Etheostoma lawrencei*
- Highland Rim darter, *Etheostoma kantuckeense*
- Holiday darter, *Etheostoma brevirostrum*
- Iowa darter, *Etheostoma exile*
- Johnny darter, *Etheostoma nigrum*
- Kanawha darter, *Etheostoma kanawhae*
- Kentucky darter, *Etheostoma rafinesquei*
- Least darter, *Etheostoma microperca*
- Lipstick darter, *Etheostoma chuckwachatte*
- Lollypop darter, *Etheostoma neopterum*
- Longfin darter, *Etheostoma longimanum*
- Maryland darter, *Etheostoma sellare*
- Mexican darter, *Etheostoma pottsii*
- Missouri saddled darter, *Etheostoma tetrazonum*
- Mud darter, *Etheostoma asprigene*
- Niangua darter, *Etheostoma nianguae*
- Okaloosa darter, *Etheostoma okaloosae*
- Orangebelly darter, *Etheostoma radiosum*
- Orangefin darter, *Etheostoma bellum*
- Orangethroat darter, *Etheostoma spectabile*
- Paleback darter, *Etheostoma pallididorsum*
- Pinewoods darter, *Etheostoma mariae*
- Rainbow darter, *Etheostoma caeruleum*
- Redband darter, *Etheostoma luteovinctum*
- Redfin darter, *Etheostoma whipplei*

- Redline darter, *Etheostoma rufilineatum*
- Redspot darter, *Etheostoma artesiae*
- Relict darter, *Etheostoma chienense*
- Rio Grande darter, *Etheostoma grahami*
- Salado darter, *Etheostuma segrex*
- riverweed darter, *Etheostoma podostemone*
- Rock darter, *Etheostoma rupestre*
- Rush darter, *Etheostoma phytophilum*
- Saffron darter, *Etheostoma flavum*
- Savannah darter, *Etheostoma fricksium*
- Sawcheek darter, *Etheostoma serrifer*
- Seagreen darter, *Etheostoma thalassinum*
- Sharphead darter, *Etheostoma acuticeps*
- Shawnee darter, *Etheostoma tecumsehi*
- Slabrock darter, *Etheostoma smithi*
- Slackwater darter, *Etheostoma boschungi*
- Slough darter, *Etheostoma gracile*
- Smallscale darter, *Etheostoma microlepidum*
- Snubnose darter, *Etheostoma simoterum*
- Sooty darter, *Etheostoma olivaceum*
- Speckled darter, *Etheostoma stigmaeum*
- Splendid darter, *Etheostoma barrenense*
- Spottail darter, *Etheostoma squamiceps*
- Spotted darter, *Etheostoma maculatum*
- Stippled darter, *Etheostoma punctulatum*
- Stone darter, *Etheostoma derivativum*
- Strawberry darter, *Etheostoma fragi*
- Striated darter, *Etheostoma striatulum*
- Striped darter, *Etheostoma virgatum*
- Stripetail darter, *Etheostoma kennicotti*

- Swamp darter, *Etheostoma fusiforme*
- Swannanoa darter, *Etheostoma swannanoa*
- Tallapoosa darter, *Etheostoma tallapoosae*
- Teardrop darter, *Etheostoma barbouri*
- Tessellated darter, *Etheostoma olmstedi*
- Tippecanoe darter, *Etheostoma tippecanoe*
- Tombigbee darter, *Etheostoma lachneri*
- Trispot darter, *Etheostoma trisella*
- Tuckasegee darter, *Etheostoma gutselli*
- Tufa darter, *Etheostoma lugoi*
- Turquoise darter, *Etheostoma inscriptum*
- Tuscumbia darter, *Etheostoma tuscumbia*
- Tuskaloossa darter, *Etheostoma douglasi*
- Variegate darter, *Etheostoma variatum*
- Vermilion darter, *Etheostoma chermocki*
- Waccamaw darter, *Etheostoma perlongum*
- Warrior darter, *Etheostoma bellator*
- Watercress darter, *Etheostoma nuchale*
- Wounded darter, *Etheostoma vulneratum*
- Yazoo darter, *Etheostoma raneyi*
- Yellowcheek darter, *Etheostoma moorei*
- Yoke darter, *Etheostoma juliae*

Dartfish

Dartfishes are a family, Ptereleotridae, of goby-like fishes in the order Perciformes.

The dartfishes were formerly classified as the subfamily Ptereleotrinae of the wormfishes, Microdesmidae, but this classification made Microdesmidae polyphyletic, hence the elevation to a family within the suborder Gobioidei.

Species

FishBase lists 45 species in five genera:

- Genus *Aioliops*
  - — *Aioliops brachypterus* (Rennis & Hoese, 1987).
  - — *Aioliops megastigma* (Rennis & Hoese, 1987).
  - — *Aioliops novaeguineae* (Rennis & Hoese, 1987).
  - — *Aioliops tetrophthalmus* (Rennis & Hoese, 1987).
- Genus *Nemateleotris*
  - — Elegant firefish, *Nemateleotris decora* (Randall & Allen, 1973).
  - — Helfrichs' dartfish, *Nemateleotris helfrichi* (Randall & Allen, 1973).
  - — Fire goby, *Nemateleotris magnifica* (Fowler, 1938).
- Genus *Oxymetopon*
  - — *Oxymetopon compressus* (Chan, 1966).
  - — Blue-barred ribbon goby, *Oxymetopon cyanoctenosum* (Klausewitz Conde, 1981).
  - — *Oxymetopon filamentosum* (Fourmanoir, 1967).
  - — *Oxymetopon formosum* (Smith, 1931).
  - — *Oxymetopon typus* (Bleeker, 1861).
- Genus *Parioglossus*
  - — *Parioglossus aporos* (Rennis & Hoese, 1985).
  - — *Parioglossus dotui* (Tomiyama, 1958).
  - — Beautiful hover goby, *Parioglossus formosus* (Smith, 1931).
  - — *Parioglossus galzini* (Williams & Lecchini, 2004).
  - — *Parioglossus interruptus* (Suzuki & Senou, 1994).
  - — Lined hover goby, *Parioglossus lineatus* (Rennis & Hoese, 1985).
  - — Dart goby, *Parioglossus marginalis* (Rennis & Hoese, 1985).
  - — *Parioglossus multiradiatus* (Keith, Bosc & Valade, 2004).
  - — *Parioglossus neocaledonicus* (Dingerkus & Seret, 1992).

— Naked hover goby, *Parioglossus nudus* (Rennis & Hoese, 1985).

— Borneo hoverer, *Parioglossus palustris* (Herre, 1945).

— Philippine dartfish, *Parioglossus philippinus* (Herre, 1945).

— Rainford's dartfish, *Parioglossus rainfordi* (McCulloch, 1921).

— Rao's hover goby, *Parioglossus raoi* (Herre, 1939).

— *Parioglossus sinensis* (Zhong, 1994).

— Taeniatus dartfish, *Parioglossus taeniatus* (Regan, 1912).

— *Parioglossus triquetrus* (Rennis & Hoese, 1985).

— Vertical hover goby, *Parioglossus verticalis* (Rennis & Hoese, 1985).

- Genus *Ptereleotris*

— *Ptereleotris arabica* (Randall & Hoese, 1985).

— Blue goby, *Ptereleotris calliura* (Jordan & Gilbert, 1882).

— Panamic dartfish, *Ptereleotris carinata* (Bussing, 2001).

— Blackfin dartfish, *Ptereleotris evides* (Jordan & Hubbs, 1925).

— Lined dartfish, *Ptereleotris grammica* (Randall & Lubbock, 1982).

— Blue hana goby, *Ptereleotris hanae* (Jordan & Snyder, 1901).

— Hovering goby, *Ptereleotris helenae* (Randall, 1967).

— Blacktail goby, *Ptereleotris heteroptera* (Bleeker, 1855).

— Sad glider, *Ptereleotris lineopinnis* (Fowler, 1935).

— *Ptereleotris melanopogon* (Randall & Hoese, 1985).

— Blue gudgeon, *Ptereleotris microlepis* (Bleeker, 1856).

— *Ptereleotris monoptera* (Randall & Hoese, 1985).

— *Ptereleotris randalli* (Gasparini, Rocha & Floeter, 2001).

— Flagtail dartfish, *Ptereleotris uroditaenia* (Randall & Hoese, 1985).

— Chinese zebra goby, *Ptereleotris zebra* (Fowler, 1938).

Death Valley Pupfish

The Death Valley pupfish, *Cyprinodon salinus salinus,* is a species of fish that is the last known survivor of what is thought to have been a large ecosystem of fish species that lived in Lake Manly which dried up at the end of the last ice age leaving the present day Death Valley in California. The pupfish is adapted to the shallow, hot, salty water of a particular part of Salt Creek that flows above ground year-round, and is also sometimes referred to as Salt Creek Pupfish. A subspecies lives in nearby Cottonwood Marsh.

Subspecies

- Cottonball Marsh pupfish, *Cyprinodon salinus milleri,* threatened

Found in Cottonball Marsh on the west side of central Death Valley.

Other Cyprinodons in the Area

- Saratoga pupfish, *Cyprinodon nevadensis nevadensis*

Found at Saratoga Springs at the south end of Death Valley.

- Amargosa pupfish, *Cyprinodon nevadensis amargosa*

Found in the Amargosa River northwest of Saratoga Springs.

- Devil's Hole pupfish, *Cyprinodon diabolis,* endangered

Found in Devil's Hole 37 miles (60 km) east of Furnace Creek, in western Nevada.

Deep-water Cardinalfish

Deep-water cardinalfishes are perciform fishes in the family Epigonidae.

They are small fishes: the largest species, *Epigonus telescopes,* reaches 75 cm in length, and most species grow to no more than 20 cm or so.

They are found in temperate and tropical oceans throughout the world. They are bathydemersal fishes (inhabiting deep waters close to the sea bed) and have been found at depths down to 3,000 m.

Species

There are 33 species in seven genera:

- Genus *Brephostoma*
  - — *Brephostoma carpenteri* (Alcock, 1889).
- Genus *Brinkmannella*
  - — *Brinkmannella elongata* (Parr, 1933).
- Genus *Epigonus*
  - — *Epigonus affinis* (Parin & Abramov, 1986).
  - — *Epigonus angustifrons* (Abramov & Manilo, 1987).
  - — *Epigonus atherinoides* (Gilbert, 1905).
  - — *Epigonus constanciae* (Giglioli, 1880).
  - — *Epigonus crassicaudus* (de Buen, 1959).
  - — *Epigonus ctenolepis* (Mochizuki & Shirakihara, 1983).
  - — Pencil cardinal, *Epigonus denticulatus* (Dieuzeide, 1950).
  - — *Epigonus devaneyi* (Gon, 1985).
  - — *Epigonus elegans* (Parin & Abramov, 1986).
  - — *Epigonus elongatus* (Parr, 1933).
  - — *Epigonus fragilis* (Jordan & Jordan, 1922).
  - — *Epigonus glossodontus* (Gon, 1985).
  - — *Epigonus heracleus* (Parin & Abramov, 1986).
  - — Big-eyed cardinalfish, *Epigonus lenimen* (Whitley, 1935).

Big-eyed Cardinalfish, Epigonus Lenimen

— *Epigonus macrops* (Brauer, 1906).
— *Epigonus marimonticolus* (Parin & Abramov, 1986).
— *Epigonus merleni* (McCosker & Long, 1997).
— *Epigonus notacanthus* (Parin & Abramov, 1986).
— *Epigonus occidentalis* (Goode & Bean, 1896).
— *Epigonus oligolepis* (Mayer, 1974).
— Bigeye, *Epigonus pandionis* (Goode & Bean, 1881).
— *Epigonus parini* (Abramov, 1987).
— *Epigonus pectinifer* (Mayer, 1974).
— Robust cardinalfish, *Epigonus robustus* (Mead & De Falla, 1965).
— Bulls-eye, *Epigonus telescopus* (Risso, 1810).
— *Epigonus waltersensis* (Parin & Abramov, 1986).

- Genus *Florenciella*
  — *Florenciella lugubris* (Mead & De Falla, 1965).
- Genus *Microichthys*
  — *Microichthys coccoi* (Ruppell, 1852).
  — *Microichthys sanzoi* (Sparta, 1950).
- Genus *Rosenblattia*
  — *Rosenblattia robusta* (Mead & De Falla, 1965).
- Genus *Sphyraenops*
  — *Sphyraenops bairdianus* (Poey, 1861).

Deep-water Stingray

The Deep-water stingray, *Plesiobatis daviesi,* is a species of ray, the only species in the genus *Plesiobatis* and family Plesiobatidae.

It is native to the Indian and western Pacific Oceans, from South Africa to Japan and Australia. It inhabits the bottoms of muddy and sandy continental slopes, where it eats small fish, crustaceans, cephalopods, and polychaetes It grows to 2.7 m in diameter.

Delta Smelt

Delta smelt, *Hypomesus transpacificus*, are slender-bodied smelts, about 5 to 7 cm long, of the Osmeridae family. They have a steely blue sheen on the sides and seem almost translucent. Smelts live together in schools and feed on zooplankton (small fishes and invertebrates). One female may lay from 1,400 to 1,800 eggs. Mature unfertilized eggs are about 1 mm.

Habitat

Delta smelt are endemic to the Sacramento Delta, California, where it is distributed from the Suisun Bay upstream through the Delta in Contra Costa, Sacramento, San Joaquin, Solano and Yolo counties.

The delta smelt is a pelagic (live in the open water column away from the bottom) and euryhaline species (tolerant of a wide salinity range). They have been collected from estuarine waters up to 14 ppt (parts per thousand) salinity.

Life Cycle

Most delta smelt live one year and die after their first spawning (semelparous). Delta smelt spawning occurs in spring in river channels and tidally influenced backwater sloughs upstream of the mixing zone (saltwater-freshwater interface). The Sacramento and San Joaquin rivers then transport the delta smelt larvae downstream to the mixing zone, normally located in the Suisun Bay.

Young delta smelt then feed and grow in the mixing zone before starting their upstream spawning migration in late fall or early winter.

Delta smelt used to be a common fish in the Sacramento-San Joaquin rivers estuary. As at 2006, however, the population is much smaller than historically and the species is now listed as threatened under the Federal Endangered Species Act (*Federal Register* 58:12863; March 5, 1993). Critical habitat was listed for delta smelt on December 19, 1994 (*Federal Register* 59:65256).

Pomacentridae

Pomacentridae (New Latin, from Greek "poma" operculum + "kentron" sting) is a family of perciform fish, comprising the damselfishes and clownfishes. They are exclusively marine (rarely brackish), and noted for their hardy constitutions and territoriality. Many are brightly coloured, so they are popular in aquaria.

Around 360 species are classified in this family, in approximately 28 genera. Of these, members of two genera, *Amphiprion* and *Premnas* are commonly called clownfish or anemonefish, while members of other genera (e.g. *Chromis*) are commonly called damselfish.

Denticle Herring

The denticle herring (*Denticeps clupeoides*) is a small (15 cm) species of ray-finned fish found only in the rivers of Benin, Nigeria and Cameroon, related to the herrings, but notable for its large anal fin and its array of denticle like scales under the head giving it almost a furry appearance. It is the sole living member of the family Denticipitidae.

Devario

The genus *Devario* comprise of some danionins familiar to aquarists. Generally (but not all) larger fish than *danio*s, they have short barbels (if present at all) and generally have deeper bodies than *danio*s with species having vertical stripes present (as well as horizontal). In size they range from 5 cm/ 2 in) to 15 cm/ 6 in).

Characteristics

They are native to the fresh water rivers and streams of South East Asia, but many species are brightly coloured, and are available as aquarium fish worldwide. A number of the species have only been recently discovered, in remote inland areas of Laos and Myanmar and do not yet have scientific names.

They generally do not live for more than about two to three years and are probably annual fish in the wild although the larger species may live for up to 5 years.

In the wild, these fish consume various small aquatic insects, crustaceans, worms, as well as, in the case of fry plankton.

Aquarium Care

The care of members of the genus *Devario* are rather similar and easily generalised. They are generally easy to moderate in difficulty to keep.

All of these fish are primarily surface feeders. They are omnivorous in the aquarium and will accept a wide variety of foods, though flake food is appropriate. Living in aquaria, live/frozen flaked foods are suitable, especially brine shrimp and sinking tablets. When conditioning danios for breeding, it's advisable to feed them plenty of fresh foods.

Although boisterous and liable to chase each other and other fish, they are good community fish and will not generally attack each other or other fish, although they occasionally nip fins, more by accident than design and will, like most fish, eat eggs and any fish small enough to fit into their mouths, in the case of the larger devarios this could include fish such as small Tetras

They are best kept in a tank long enough for their active swimming, preferably with a current from a power filter (or at least airstone) as they often live in fast flowing streams in the wild. Generally this also results in them being sub tropical with cooler temperatures. They are good jumpers and a tight fitting lid is recommended

Schooling fish, they prefer to be in groups of six or more. Devarios prefer water with a 6.0-8.0 pH, a water hardness of up to 5.0-19.0 dGH, and a temperature range of 68-80 °F (18-24 °C), the lower end of the temperature range is ideal.

Breeding

Some species of *Devario,* such as the Giant danio are among the easiest aquarium fish to breed. Other species such as Devario pathirana are far harder to spawn. All scatter their eggs over the substrate. The eggs are adhesive and hatch within 2-3 days. Eggs will be eaten enthusiastically unless protected by a heavy planting, e.g. Java moss or a spawning mop.

Danios and Devarios

Most species that were formerly within the *danio* genus such as the Giant danio have now been reclassified into the *Devario* genus. For full details of the classification of *danios* and *Devarios* are on the danionin page.

Discus

Discus (*Symphysodon* spp.) are a genus of three species of freshwater cichlid fishes native to the Amazon River basin. Discus are popular in aquarium fish and their aquaculture in several countries in Asia is a major industry.

Taxonomy

Discus belong to the genus Symphysodon, which currently includes three species: The common discus (*Symphysodon aequifasciatus*), the Heckel discus (*Symphysodon discus*), and a new species which has been named *Symphysodon tarzoo.*

Appearance

Like cichlids from the genus *Pterophyllum,* all *Symphysodon* species have a laterally compressed body shape. In contrast to *Pterophyllum,* however, extended finnage is absent giving *Symphysodon* a more rounded shape. It is this body shape from which their common name, "discus" is derived. The sides of the fish are frequently patterned in shades of green, red, brown, and blue. The height and length of the grown fish are both about 20–25 cm (8–10 in).

Reproduction and Sexual Dimorphism

Another characteristic of *Symphysodon* species are their care for the larvae. As for most cichlids, brood care is highly developed with both the parents caring for the young. Additionally, adult discus produce a secretion through their skin, off which the larvae live during their first few days. This behaviour has also been observed for *Uaru* species.

Diet

In the wild they are opportunistic omnivores and their diet consists of invertebrates and plants. The waters from which discus hail are typically slow-moving, soft and slightly acidic (1 - 5 dGH, pH 4.0 – 6.5). Temperature of the water in their natural habitat varies from 25 – 30 C (82-86 F). The males penis is around 3mm long making mating with a female very easy.

Distribution

The three species of *Symphysodon* have different geographic distributions. *S. aequifasciatus* occurs in the Rio Solimões, Rio Amazonas and the Río Putumayo-Içá in Brazil, Colombia and Peru. In contrast the distribution of *S. discus* appears to be limited to the lower reaches of the Abacaxis, Rio Negro and Trombetas rivers. *S. tarzoo* occurs upstream of Manaus in the western Amazon.

In the Aquarium

Discus are shy and generally peaceful aquarium inhabitants. They are sensitive to stress and disturbance or lack of protection. The best cohabitants may be angelfish (although some aquarists claim that keeping them together with angelfish will introduce parasites and/or diseases) and small characides like tetras. Uaru species are also suggested cohabitants for discus. It is noteworthy, however, that small fish may be intimidated or eaten by the discus. Catfish with sucker mouths are less than ideal cohabitants for discus since they sometimes attach themselves on the sides of discus and eat their mucus membranes.

Many aquarists consider discus to be finicky and not particularly hardy. They often become susceptible to disease and die if not kept in optimal conditions.

Aquarium Water Chemistry

Aquariums for discus should be kept within a temperature range of 26-31°C (82-86°F); a temperature of 29°C (84°F) is thought ideal for adults. Babies and young fish should be maintained at 31°C (86°F) degrees. The water should be very soft and slightly acidic; a pH of 5.5 - 6.5 is considered good for wild caught discus.

Captive bred fish adapt very well to harder water and to pH up to 7.2, except when attempting to breed, in which case soft and acidic is best, although it is preferred by the fish anyway. VERY clean water with frequent large volume water changes is necessary for the health of these fish.

Never use pure R.O. water or distilled water as some "salts" are necessary. (ie;calcium,magnesium, etc.) 100 ppm GH is average. New fish should be quarantined for a minimum of 4-6 weeks in a separate room, separate tank, and separate water changing equipment to eliminate the possibility of bringing in an infection to established fish. It is generally accepted that new fish should be added after "lights out" or during normal feeding.

Water quality must be very high, as discus do not tolerate pollution of any sort very well. A good tank will be equipped with a high capacity biological filter and be fully cycled (which usually takes a month or more.) Ammonia and nitrites should be kept at 0 ppm. Nitrates should also be kept as low as possible. Weekly water changes are important, except in the case of a very heavily planted tank with high nitrogen compound grounding capacity and a very small biological load.

Feeding

Feeding discus is sometimes a challenge. They have no unique nutritional requirements; they can be raised on just about any high-protein fish food. However, discus are often extremely cautious about new foods; it is not unusual for them to go for weeks without food before accepting a new type of food. (Therefore, when purchasing discus it is a good idea to ask what they are being fed.) After starving for a month discus will almost always accept a new food, but this may stunt the growth of younger fish.

It is not advisable to use the starving method for weaning discus off of one food for another. Instead, mix the new food with the discus' preferred food. Over time, the discus will begin to accept the new food, and the old can be removed.

Beef heart is often fed to discus in order to promote good colouration and quick growth. However, concern over the long-term consequences of feeding discus a diet high in mammalian protein has prompted some hobbyists to switch their discus to a diet of krill, a shrimp-like crustacean. You can also feed them discus dinner which is a mixture of all the proteins and nutrients that discus need to be perfectly healthy.

Lighting

Discus prefer low lighting. They are often skittish in the home aquarium, so low lighting together with profuse aquatic vegetation may help them to feel more comfortable in their environment.

Common Colour Varieties

There are Three Layers of Colour on Discus: The base colour (which usually ranges from cream to red-brown), the secondary colour (a metallic colour, usually a blue or green colour) and the black pigment that makes up the black vertical bars and allows the fish to darken and lighten at will.

Most discus strains have either a golden or reddish base colour. The secondary colour is often striped down the sides of the fish, although many strains (such as 'solid cobalt' or 'blue diamonds') have secondary colour that eventually covers most or all of the fish's body.

There are no rules or authorities on what constitutes a unique colour variety or what to call it. A particular form may or may not breed 'true' (with offspring very closely resembling the patterns of their parents.) Generally all of the common, established forms breed true. The exact patterning of the secondary (blue/green) colour is like a fingerprint; it develops chemically rather than being set precisely by genetics. The offspring of two 'spotted' discus will likely have spots, but not in the exact same size/position as their parents.

Notable Colour Varieties:

- *Brown:* The most common colour form in the wild; these fish have a brownish base colour with minimal stripes of secondary colour only along the head and fins.
- *Blue/Green:* Similar to the Brown, but with more secondary colour (either bluish or greenish.)
- *Royal Blue:* The secondary colour forms stripes across the entire body, with a golden base colour. These splendid fish are the basis of many of the developed colour strains, and are primarily responsible for the early fame of discus. Royal Blues can usually be readily distinguished from selectively bred colour forms by their less even base colour, with the golden colour becoming a brighter yellow around the breast area.
- *Red Spotted Green:* A reddish base colour with greenish secondary colour with 'holes' in it (producing spots of the red base colour showing through.) This handsome colour form is extremely rare in the wild, but is produced by several breeders.
- *Heckel:* Possibly a separate species, Heckels are identifiable by two vertical black bars that are much thicker than the others.

Common Bred forms:

- *Red Turquoise:* A red-brown base colour with stripes of blue-green secondary colour, normal black pigmentation (bars).
- *Solid Cobalt:* Golden or light brown base colour, but when fully mature covered with a blue secondary colour. Black pigmentation may be normal or incomplete (some vertical bars missing.)
- *Blue Diamond:* Essentially a 'solid cobalt', but the black bars have been completely removed through selective breeding. The reduction in black pigment gives these fish a bright, lighter blue colour than most 'solid' discus.

- *The Pigeon Blood Mutants:* These fish have a gene that disrupts the distribution of the black pigment. As a result, they lack vertical black bars (but often have 'pepper'). The lack of black pigment makes their base colour much lighter and brighter; as a result, discus with this mutation may show brilliant red or yellow (or even pale cream) primary colour. Most of these strains are no longer called 'pigeon bloods' per se, but are easily identifiable by the bright base colour, pepper, and lack of black vertical bars. All pigeon bloods are the descendant of a single fish found in Eastern Asia in the 1980s. Since the trait is dominant and appears to be controlled by a single gene, fish bearing this mutation can be crossed with any other colour strain to produce novel new 'pigeon blood' types. Pigeon bloods do have one drawback: They cannot darken at will (as normal discus can). This can make it difficult for them to raise fry, which are attracted to their parents by seeking out a dark object. (Normal discus darken when spawning or stressed.) The fish shown at the top of this document is a pigeon blood. (High quality pigeon blood types have few or no 'pepper'.)
- *The Extremley Rare Glass Discus:* These Fish are transparent apart from the internal organs. These fish are found in the deep amazon, where they are protected by the local tribe. These are the shyish discus of them all.
- *Snake-skins:* These fish have a mutation that makes their patterning 'tighter'; as a result, they have about twice as many black vertical bars, but also have tighter, finer secondary colour patterns than normal discus.

3

Freshwater Halfbeaks

The freshwater halfbeaks of the genera *Dermogenys*, *Hemirhamphodon*, and *Nomorhamphus* are all livebearers, that is, they do not lay eggs but instead produce well-developed free-swimming young. However, there is a great deal of variation in the details. Meisner and Burns identified no fewer than five distinct modes of viviparity and ovoviviparity in the freshwater halfbeaks:

- *Type 1:* Fertilized eggs retained within the ovarian follicle. Superfetation, that is storage of sperm, does not occur. The eggs are provided with a large yolk sac and have little or no connection to the maternal blood supply. Example: South East Asian populations of *Dermogenys pusilla*.
- *Type 2:* Fertilized eggs retained within the ovarian follicle. Superfetation does occur, with up to three broods resulting from a single mating. The eggs are provided with a small yolk sac but the embryos instead have a connection to the maternal supply through the coelomic cavity and pericardial sac. Examples: *Dermogenys pusilla* from Sabah and *Dermogenys orientalis*.

- *Type 3:* Fertilized eggs retained within the ovarian follicle only for the early stages of development, with the embryos later developing along the full length of the ovary. Superfetation does occur, and up to two broods can develop simultaneously in the ovary. The eggs are provided with a small yolk sac but the embryos have a connection to the maternal supply through an expanded belly sac. Example: *Dermogenys viviparus.*
- *Type 4:* Fertilized eggs retained within the ovarian follicle only for the early stages of development, with the embryos later developing along the full length of the ovary. Superfetation does not occur. The eggs are provided with a large yolk sac and the embryos have no connection to the maternal blood supply. Examples: *Nomorhamphus megarrhamphus, Nomorhamphus weberi,* and *Nomorhamphus towoetii.*
- *Type 5:* Fertilized eggs retained within the ovarian follicle only for the early stages of development, with the embryos later developing along the full length of the ovary. Superfetation does occur, and embryos of different ages can be found in the ovaries. The eggs are provided with a small yolk sac and the embryos only have a connection to the maternal blood supply for only part of their development. Late-stage embryos appear to eat eggs and small embryos in ovary. Note that embryos eating eggs and other embryos has been observed in a few other fishes, most notably sharks. Example: *Nomorhamphus ebrardtii.*

As with other livebearing fish, freshwater halfbeaks produce small broods of large offspring compared with egg-laying species of similar size, with broods of around ten to twenty, 10-15 mm long offspring being typical.

Sexual Dimorphism

Sexual dimorphism is apparent in some species of halfbeak. Males of the ovovivaparous and vivaparous species all have a modified anal fin, the andropodium, similar to the gonopodium of poecilid livebearers, used to deliver sperm to the females.

Although most of the egg laying species mate by shedding the milt externally, as is typical for bony fish, at least some egg-laying species practise internal fertilization: male *Zenarchopterus* use a modified anal fin to direct sperm into the genital opening of the female prior to spawning . Halfbeaks therefore, show the full range of reproductive methods know to occur in fishes: external fertilization with egg laying; internal fertilization with egg laying; internal fertilization with ovoviviparity; and internal fertilization with viviparity.

Besides modifications to the anal fin, sexual dimorphism is quite strong in some species, especially among the livebearing freshwater halfbeaks. Female *Normorhamphus* are much larger than males but aren't as brightly coloured and have shorter beaks. By contrast, male *Hemirhamphodon* are larger than the females, and some species, such as *Hemirhamphodon pogonognathus*, also have a long beard-like tassle on the end of the beak.

Halfbeaks in Aquaria

Species of the genera *Dermogenys* and *Nomorhamphus* are quite commonly kept as aquarium fish; species of *Hemirhamphodon* and *Zenarchopterus* are rather less commonly seen. They are small and generally peaceful towards other species, although males can be aggressive to one another.

Male *Dermogenys pusillius* in particular fight vigorously and sometimes these battles end in injuries; this fish has therefore become known as the wrestling halfbeak and in some Asian countries fights between males are used for betting purposes in much the same way as the Siamese fighting fish.

To be kept successfully, halfbeaks require an aquarium with plenty of space at the surface. Depth is not critical, so a tank that is wide is better than one that is deep. They are sensitive to low oxygen levels but are otherwise relatively hardy, with one important exception: they do not tolerate sudden changes in salinity, pH, hardness, or temperature well. They must be introduced to a new aquarium gently, and subsequent water changes are best small but frequent so that the water chemistry does not suddenly change.

A few species, most notably *Dermogenys pusillius*, have traditionally been said to do best in slightly brackish water though some authors disputer this, suggesting that reports that these fish come from brackish water have come about because juvenile estuarine and marine halfbeaks have been misidentified as adult freshwater halbeaks. Most of the traded species of *Nomorhamphus* and *Hemirhamphodon* are known to prefer soft, neutral to slightly acid, freshwater conditions.

Halfbeaks are nervous fish and things like switching on lights can cause them to swim around the tank frantically. They may hit themselves on the glass, injuring their beaks, or jump out of the tank completely. Injuries to the beak usually heal within a few weeks. They will eat insect larvae such as bloodworms readily, as well as crustacean eggs, shrimps, fruit flies, and even small pieces of chopped white fish. Halfbeaks sometimes eat flake foods as well. Some aquarists also offer them tiny pieces of algae wafer on the basis that most species are omnivorous in the wild, and so a certain amount of green food probably does them good.

Halfbeaks will breed in captivity, but despite being livebearers, they are not particularly easy to breed. Miscarriages are common, particularly if the females are stressed or shocked (for example, by being moved to another aquarium). Once the fry have been born, things get much simpler, as the baby halfbeaks are quite big and will eat newly hatched brine shrimps, small live foods such as daphnia, and powdered flake.

Halibut

A halibut is a type of flatfish from the family of the righteye flounders (Pleuronectidae). This name is derived from Dutch *heilbot*. Halibut live in both the North Pacific and the North Atlantic oceans, and are highly regarded food fish.

Physical Characteristics

The Halibut is the largest of all flatfish, with an average weight of about 25 lb - 30 lb, but they can grow to be as much as 600 lbs. The Halibut is blackish-grey on the top side and off-white on the underbelly side.

When the Halibut is born the eyes are on both sides of its head so it has to swim like a salmon. After about 6 months one eye will rotate to the other side of its head, making it look more like the flounder . This happens at the same time that the stationary eyed side begins to develop a blackish-grey pigment while the other side remains white. This disguises a halibut from above (blending with the ocean floor) and from below (blending into the light from the sky).

Diet

Halibut feed on almost any animal they can fit in their mouths. Animals found in their stomachs include sand lance, octopus, crab, salmon, hermit crabs, lamprey, sculpin, cod, pollock, herring and flounder.

Halibut can be found at depths as shallow as a few metres to hundreds of metres deep, and although they spend most of their time near the bottom, halibut will move up in the water column to feed. In most ecosystems the halibut is near the top of the marine food chain. In the North Pacific the only common predators of halibut are the sea lion (*Eumetopias jubatus*), the orca whale (*Orcinus orca*), and the salmon shark (*Lamna ditropis*).

Halibut Fishery

The commercial halibut fishery in the North Pacific dates to the late 19th century and today is one of the largest and most lucrative fisheries in the region. In Canadian and US waters of the North Pacific, halibut are taken by longline, using chunks of octopus ("devilfish") or other bait on circle hooks attached at regular intervals to a weighted line that can extend for several miles across the bottom. Typically, the fishing vessel hauls gear after several hours up to a day has passed.

Careful international management of Pacific halibut is necessary, as the species occupies the waters of the United States, Canada, Russia, and possibly Japan, and is a slow-maturing fish. Halibut do not reproduce until age eight, when they are approximately 30 inches (76 cm) long, so commercial capture of fish below this length is an unsustainable practice and is against US and Canadian regulations. The halibut fishery in the Pacific is managed by the International Pacific Halibut Commission (IPHC).

For most of its modern duration, the commercial halibut fishery operated as a derby-style fishery where regulators declared time slots when the fishery was open (typically 24-48 hours at a time) and fisherman raced to catch as many pounds as they could within that window.

This approach accommodated unlimited participation in the fishery, while allowing regulators to control the quantity of fish caught annually by controlling the number and timing of openings. The approach frequently led to unsafe fishing as openings necessarily set in advance and fisherman felt compelled economically to leave port virtually regardless of the weather. The approach also provided fresh halibut to the markets for only several weeks each year.

In 1995, regulators in the United States implemented a quota-based fishery by allocating individual fishing quotas (IFQs) to existing fishery participants based on each vessel's documented historical catch. IFQs grant holders a specific proportion of each year's total allowable catch (TAC) as determined by regulators, and can be fished at any time during the 9-month open season.

The IFQ system improved both the safety of the fishery and the quality of the product by providing a stable flow of fresh halibut to the marketplace. Critics of the programme suggest that, since IFQs are a saleable commodity and the fish a public resource, the IFQ system gave a public resource to the private sector. Would be fisherman, who were not part of the initial IFQ allocation are also critical of the programme saying that the capital costs to fishery entry are now too high.

There is also a significant sport fishery in Alaska and British Columbia where halibut are a prized game and food fish. Sport fishermen use large rods and reels with line weights from 80 to 150 pound test, and often bait with herring, large jigs, or even whole salmon heads. Halibut are very strong, thus in both commercial and sport fisheries large halibut (over 50 to 100 pounds (20 to 50 kg)) are often shot or otherwise subdued before they are brought onto the boat.

The sport fishery in Alaska is one of the key elements to the state's summer tourism economy. It is to be noted however, that the amount of halibut caught sport fishing is considerable and fish are frequently below 30" inches in length. The likelihood of overfishing and unenforcible regulations has prompted the government to consider a moratorium on sport fishing for halibut. Halibut are typically broiled, deep fat fried or lightly grilled while fresh.

The filets can also be smoked but this method is more difficult with halibut meat than it is with salmon, due to the ultra-low fat content of halibut. Eaten fresh, the meat has a very clean taste and requires little seasoning. Halibut is also noted for its very dense and firm texture, almost more akin to chicken.

Halibut have been an important food source to Native Americans and Canadian First Nations for thousands of years and continue to be a key element to many coastal subsistence economies. The management of the halibut resource to accommodate the competing interests of commercial, sport, and subsistence users is a contentious current issue.

The Atlantic Fishery of halibut has been extremely depleted through overfishing to such an extent that it may possibly be declared an endangered species. Almost all halibut now bought on the East coast are now Pacific halibut.

Species Commonly known as "Halibut"

- Family Carangidae (jack family, not a flatfish)
  - — Australian halibut, *Parastromateus niger*
- Family Paralichthyidae
  - — California halibut, *Paralichthys californicus*
  - — Bastard halibut, *Paralichthys olivaceus*
- Family Pleuronectidae
  - — Arrowtooth halibut, *Atheresthes evermanni*
  - — Shotted halibut, *Eopsetta grigorjewi*
  - — Atlantic halibut, *Hippoglossus hippoglossus*
  - — Pacific halibut, *Hippoglossus stenolepis*
  - — Greenland halibut, *Reinhardtius hippoglossoides*
  - — Spotted halibut, *Verasper variegatus*
- Family Psettodidae
  - — Indian halibut, *Prettads erumei*

Halosaur

Halosaurs are eel-shaped fishes found only at great ocean depths. As the family Halosauridae, halosaurs are one of two families within the order Notacanthiformes; the other being the deep-sea spiny eels. Halosaurs are thought to have a worldwide distribution, with some seventeen species in three genera represented. Only a handful of specimens have been observed alive, all via chance encounters with remotely operated submersibles.

From the Greek *hals* meaning "sea" and *sauros* meaning "lizard", halosaurs look like living fossils from some throwback era. Their greatly elongated bodies end in a whip-like tail; their scales are large. There is one small dorsal fin close to the sharply pointed, mostly scaleless head. The tail fin is greatly reduced, with the anal fin being the largest fin. Their pecotral fins are slender and also greatly elongated. The mouth is somewhat large, with the lower jaw shorter than the upper jaw. The gas bladder is absent.

The largest species, the 90 centimetre long abyssal halosaur (*Halosauropsis macrochir*) is also one of the most deep-living fish, recorded at depths of 3,300 metres. Halosaurs have developed certain adaptations to life at these extreme depths, where no light penetrates. Their lateral line system is highly developed; this is a system of pores running the length of the fish's body, lending it a sort of "sixth sense" by detecting nearby vibrations. Some species are also known to hold their elongate pectorals erect and forward, possibly providing a further means of detection.

Halosaurs are benthic fish, spending their time cruising over or resting on the sea floor where temperatures may be just 2-4 degrees Celsius. They propel themselves with rhythmic undulations of the body, not unlike snakes. Halosaurs are thought to prey mainly on benthic invertebrates, such as polychaete worms, echinoderms and crustaceans such as copepods.

In life, most halosaurs are a grey to bluish black in colour. Like other notacanthiform fish, halosaurs are able to regenerate their tails easily if lost. This adaptation can be likened to certain terrestrial reptiles such as the glass lizard, which sacrifices its tail in order to evade predators.

Species

There are seventeen species in three genera:

- Genus *Aldrovandia*
  - — Gilbert's halosaur, *Aldrovandia affinis* (Gunther, 1877).
  - — *Aldrovandia gracilis* (Goode & Bean, 1896).
  - — *Aldrovandia mediorostris* (Gunther, 1887).
  - — *Aldrovandia oleosa* (Sulak, 1977).
  - — Hawaiian halosaur, *Aldrovandia phalacra* (Vaillant, 1888).
  - — *Aldrovandia rostrata* (Gunther, 1878).
- Genus *Halosauropsis*
  - — Abyssal halosaur, *Halosauropsis macrochir* (Gunther, 1878).

- Genus *Halosaurus*
  - — *Halosaurus attenuatus* (Garman, 1899).
  - — *Halosaurus carinicauda* (Alcock, 1889).
  - — *Halosaurus guentheri* (Goode & Bean, 1896).
  - — *Halosaurus johnsonianus* (Vaillant, 1888).
  - — *Halosaurus ovenii* (Johnson, 1864).
  - — *Halosaurus parvipennis* (Alcock, 1892).
  - — Goanna fish, *Halosaurus pectoralis* (McCulloch, 1926).
  - — *Halosaurus radiatus* (Garman, 1899).
  - — *Halosaurus ridgwayi* (Fowler, 1934).
  - — *Halosaurus sinensis* (Abe, 1974).

Hamlet

A hamlet is a fish of the genus *Hypoplectrus*. It is a grouping fish that is found mainly in coral reefs in the Caribbean Sea and the Gulf of Mexico, particularly around Florida and the Bahamas.

Reproduction

Hamlets are *simultaneous hermaphrodites* (or *synchronous hermaphrodites*): They have both male and female sexual organs at the same time as an adult. They seem quite at ease mating in front of divers, allowing observations in the wild to occur readily. They do not practice self-fertilization, but when they find a mate, the pair takes turns between which one acts as the male and which acts as the female through multiple matings, usually over the course of several nights.

Hammerhead Shark

Hammerhead sharks of the genus *Sphyrna* are members of the family Sphyrnidae. The only other genus of Sphyrnidae, *Eusphyra*, contains only one species, *Eusphyra blochii*, the winghead shark.

The nine known species of hammerhead range from 2 to 6 m long (6.5 to 20 feet), and all species have a projection on each side of the head that give it a resemblance to a flattened hammer. The shark's eyes and nostrils are at the tips of the extensions.

They are aggressive predators which eat fish, rays, cephalopods, and crustaceans. They are found in warmer waters along coastlines and continental shelves.

The hammer shape of the head was once thought to act as a wing, aiding in close-quarters manoeuvrability and allowing the shark to execute sharp turns without loss of stability. However, it was found that the special design of its vertebra allowed it to make the turns correctly, more than its head. But as a wing the hammer would also provide lift; hammerheads are one of the most negatively buoyant of sharks. Like all sharks, hammerheads have electrolocation sensory pores called ampullae of Lorenzini. By distributing the receptors over a wider area, hammerheads can sweep for prey more effectively. These sharks have been able to detect an electrical signal of half a billionth of a volt. The hammer shaped head also gives these sharks larger nasal tracts, increasing the chance of finding a particle in the water by at least 10 times as compared to other 'classical' sharks.

Wider spacing between sensory organs better enables an organism to detect gradients and therefore the location of a gradient source such as food or a mate. The peculiar head of this shark can be thought of as analogous to the antennae of an insect.

Hammerheads have proportionately small mouths and seem to do a lot of bottom-hunting. They are also known to form schools during the day, sometimes in groups of over 100. In the evening, like other sharks, they become solo hunters.

Reproduction

Reproduction in the hammerhead shark occurs once a year with each litter containing 20 to 40 pups. Hammerhead shark mating courtship is a violent affair. The male will bite the female until she acquiesces, allowing mating to occur. Unlike many other shark species, the hammerhead shark has internal fertilization which creates a safe environment for the sperm to unite with the egg. The embryo develops within the female inside a placenta and is fed through an umbilical cord, similar to mammals.

The gestation period is 10 to 12 months. Once the pups are born the parents do not stay with them and they are left to fend for themselves. A world-record 1,280 pound (580 kg) pregnant female hammerhead shark was caught off Boca Grande, Florida on May 23, 2006. The shark was carrying 55 pups, which suggests scientists had previously underestimated the number of pups per gestation.

In May 2007 scientists discovered that Hammerhead sharks can reproduce asexually through a rare method known as parthenogenesis, as they have the ability to fertilize their own eggs. At first skepticised wildly, due to the fact that a female shark can store sperm inside her for months, even years, but it was confirmed through DNA testing that the pup lacked any paternal DNA. This is the first documented case of any shark doing this.

Species

Of the nine known species of hammerhead, three can be dangerous to humans: the scalloped, great, and smooth hammerheads.

- Genus *Sphyrna*
  - Subgenus *Sphyrna*
    - ⇒ Scalloped hammerhead, *Sphyrna (Sphyrna) lewini* (Griffith & Smith, 1834)
    - ⇒ "Cryptic scalloped hammerhead"—Scalloped hammerheads turn out to be divided into two separate species, which have not been officially reclassified with separate names.
    - ⇒ Great hammerhead, *Sphyrna (Sphyrna) mokarran* (Ruppell, 1837)

⇒ Smooth hammerhead, *Sphyrna (Sphyrna) zygaena* (Linnaeus, 1758)

⇒ Whitefin hammerhead, *Sphyrna (Sphyrna) couardi* (Cadenat, 1951)

— Subgenus *Mesozygaena*

⇒ Scalloped bonnethead, *Sphyrna (Mesozygaena) corona* (Springer, 1940)

⇒ Squarehead Shark Sphyrna (*Mesozygaena*) sp. listed on elasmo-research's list

— Subgenus *Platysqualus*

⇒ Scoophead, *Sphyrna (Platysqualus) media* (Springer, 1940)

⇒ Bonnethead or shovelhead, *Sphyrna (Platysqualus) tiburo* (Linnaeus, 1758)

⇒ Smalleye hammerhead, *Sphyrna (Platysqualus) tudes* (Valenciennes, 1822)

Announcements in June, 2006 reported the discovery of a possible new species of hammerhead of the shores of South Carolina. The possible new species is referred to simply as a cryptic species until it receives an official designation. This is prolonged, in part, because the discovery is really that the "scalloped hammerhead" is possibly two different species, not that a new species has been sighted, in the normal way. The discovery that scalloped hammerheads are possibly two species is purely a result of genetic testing, not identification of physical differences.

Since sharks do not have mineralised bones and rarely fossilise, it is their teeth alone that are commonly found as fossils. The hammerheads seem closely related to the carcharhinid sharks that evolved during the mid-Tertiary Period. Because the teeth of hammerheads resemble those of some carcharhinids, it has been difficult to determine when hammerheads first appeared. It is probable that the hammerheads evolved during the late Eocene, Oligocene or early Miocene.

Geneticist Andrew Martin used DNA to study all of the hammerhead species and he concluded that the first hammer appeared on the winghead shark, which has the largest hammer, and the rest of the hammerhead sharks evolved one at a time from the original winghead shark each with a smaller hammer.

Hammerjaw

The hammerjaw or omosudid, *Omosudis lowii*, is a small deep-sea aulopiform fish, found worldwide in tropical and temperate waters to 4,000 m (13,000 ft) depth. It is the only representative of its family, Omosudidae (from the Greek *omo*, "shoulder", and Latin *sudis*, either "esox, fish of the Rhine" or "stake").

Physical Description

The large head is dominated by a massive, truncated lower jaw and large, high-set eyes. The lower jaw has a dark, almost black distal end,"chin". The lower jaw possesses at least one pair of oversized, transparent, and daggerlike teeth; the palatines possess 1–4 pairs of slightly smaller teeth. The body itself is scaleless and laterally compressed; it is covered in iridescent, silvery-gray guanine with the dark peritoneum peaking through in places.

Dark smoky gray dorsally, the body tapers strongly towards a thin caudal peduncle (which has a dermal keel) and deeply forked caudal fin. The caudal peduncle is a smoky black colour, darker than the body and tail. All fins are spineless; both the low-slung pectoral fins (with 11–16 soft rays) and abdominal pelvic fins (with 8 soft rays) are fairly small.

The single dorsal fin (with 9–12 soft rays) and anal fin (14–16 soft rays) are roughly equal in size; the anal fin's origin lies immediately posterior to the dorsal. The lateral line is uninterrupted and the gill rakers number 20–25. Like other members of their order, hammerjaws also possess a small adipose fin. The largest recorded hammerjaw measured 23 cm (9 inches) standard length (excluding the caudal fin).

Life History

Very little is known of the hammerjaw's life history. It inhabits the mesopelagic and bathypelagic zones down to 4,000 m and is never caught in large numbers.

It is inferred from their sporadic capture and sleek morphology that hammerjaws are swift swimmers—capable of avoiding sampling nets.

Hammerjaws appear to spawn year-round; like many other deep-sea aulopiform fish, they are hermaphrodites. They are carnivorous and feed on squid and other pelagic fish; in turn, hammerjaws are known prey of lancetfish and tuna.

Handfish

Handfishes are anglerfishes in the genus *Brachionichthys*, the only genus in the family Brachionichthyidae.

They are small (up to 15 cm) bottom-dwelling marine fishes found in coastal waters of southern Australia and Tasmania. Their skin is covered with denticles (tooth-like scales), giving them the name warty anglers.

They use their pectoral fins to walk about on the sea floor. These highly modified fins have the appearance of hands, hence their scientific name, from Latin *bracchium* meaning "arm" and Greek *ichthys* meaning "fish".

Like other anglerfishes, they possess an illicium, a modified dorsal fin ray above the mouth, but it is short and does not appear to be used as a fishing lure. The second dorsal spine is joined to the third by a flap of skin, making a crest.

Species

There are four species:

- Spotted handfish, *Brachionichthys hirsutus* (Lacepede, 1804).
- Red handfish, *Brachionichthys politus* (Richardson, 1844).
- *Brachionichthys unipennis* (Cuvier, 1817).
- *Brachionichthys verrucosus* (McCulloch & Waite, 1918).

Hawkfish

Hawkfish are strictly tropical, perciform marine fish of the family Cirrhitidae. Associated with the coral reefs of the western and eastern Atlantic and Indo-Pacific, the hawkfish family contains 12 genera and 32 species. They share many morphological features with the scorpionfish of the family Scorpaenidae.

Hawkfish have large heads with thick, somewhat elongate bodies. Their dorsal fins are merged, with the first consisting of ten connected spines. At the tip of each spine are several trailing filaments, hence the family name *Cirrhitidae*, from the Latin *cirrus* meaning "fringe." Their tail fins are rounded and truncate and their pectoral fins are enlarged and skinless. Their scales may be cycloid or ctenoid. Most hawkfish are small, from about 7-15 centimetres in length. The largest species, the giant hawkfish (*Cirrhitus rivulatus*) attains a length of 60 centimetres and a weight of 4 kilograms. A commercial fishery exists for the larger species as they are considered an excellent food fish.

The vibrant colours exhibited by most hawkfish have won them popularity in the aquarium hobby, aided by the fishes' reputation for unproblematic upkeep and easy acclimation to tank life. Popularly kept species include the longnose hawkfish (*Oxycirrhites typus*), coloured in a red and pink crosshatch pattern, and the flame hawkfish (*Neocirrhites armatus*).

Thanks to their large, skinless pectoral fins, hawkfish are able to perch upon flame corals without incurring harm. Actually hydrozoans rather than true corals, flame corals possess stinging cells called nematocysts, which would normally prevent close contact. Afforded some degree of protection by their living perches, hawkfish seek the high ground of the reef where they warily survey their surroundings; redolent of a hawk's behaviour, this habit inspired their common name.

Most hawkfish are solitary in nature but some will form

pairs and share a head of coral. Other species form harems of up to seven females dominated by a larger male. They are diurnal and remain within the shallows of the reef, no deeper than 30 metres. Typically motionless, hawkfish will dart out to grab crustaceans and other small invertebrates, which happen to pass by.

Spawning occurs at night, at or near the water's surface. Hawkfish are pelagic spawners; that is, they release many tiny buoyant eggs which drift with the ocean currents until hatching. Hatching is thought to happen after about three weeks; the distance travelled in this time may explain the exceptionally wide distribution of hawkfish. They have not been successfully bred in the aquarium, with the exception of the longnose hawkfish.

Hawkfish are noted for their protogynous hermaphrodism: functional females will change into males if the dominant supermale dies. Hawkfish are generally not sexually dichromatic, meaning the sexes cannot be distinguished by colouration alone.

Herring

Herrings are small oily fish of the genus *Clupea* found in the shallow, temperate waters of the North Atlantic, the Baltic Sea, the North Pacific, and the Mediterranean. There are 15 species of herring, the most abundant of which is the Atlantic herring (*Clupea harengus*). Herrings move in vast schools, coming in spring to the shores of Europe and America, where they are caught, salted and smoked in great quantities. Canned "sardines" (or pilchards) seen in supermarkets may actually be sprats or round herrings.

Morphology

All of the 200 species in the family Clupeidae share similar distinguishing features. They are silvery coloured fish that have a single dorsal fin. Unlike most other fish, they have soft dorsal fins that lack spines, though some species have pointed scales that form a serrated keel. They have no lateral line and have a protruding lower jaw. Their overall size varies from species to species: the Baltic herring is small, usually about 14 to 18 centimetres in length, the Atlantic herring can grow to about 46 cm (18 inches) in length and weigh up to 1.5 pounds (700 g), Pacific herring grow to about 38 cm (15 inches).

Predators

Predators of adult herring include seabirds, dolphins, porpoises, seals, whales, and humans. Large fish such as sharks, dog fish, tuna, cod, salmon, halibut and other large fish also feed on adult herring. Many of these organisms also prey on larval and juvenile herring.

Diet

Young herring feed on phytoplankton and as they mature they start to consume larger organisms. Adult herring feed on zooplankton, tiny animals that are found in oceanic surface waters, and small fish and fish larvae. Copepods and other tiny crustaceans are the most common zooplankton eaten by herring. During daylight herring stay in the safety of deep water, feeding at the surface only at night when there is less chance of predation. They swim along with their mouths open, filtering the plankton from the water as it passes through their gills.

Economy

Herring are an important economic fish. Adult fish are harvested for their meat and eggs. In Southeast Alaska herring is a sold as baitfish. Environmental Defence suggests Atlantic herring (*Clupea harengus*) may be the most Ecological choice for eating.

Cuisine

Herring has been a known staple food source since 3000 BC. There are numerous ways the fish is served and many regional recipes: eaten raw, fermented, pickled, or cured by other techniques. The fish was sometimes known as two-eyed steak.

Nutrition

Herring are very high in healthy long-chain Omega-3 fatty acids, EPA and DHA. They are a good source of vitamin D. They are also very low in the toxins PCBs, dioxins, and mercury.

Baltic herring slightly exceeds recommended limits with respect to dioxin.

Pickled Herring

Pickled herring is a popular traditional Scandinavian delicacy. Most home cured herring uses a two-step curing process. Initially, herring is cured with salt to extract water. The second stage involves removing the salt and adding flavourings, typically a vinegar/salt/sugar solution to which ingredients like peppercorn, bay leaves and raw onions are added.

Once the pickling process is finished and depending on which of the dozens of classic herring flavourings are selected, it is usually enjoyed with dark rye bread, crisp bread, or potatoes. This dish is a must at Christmas and Midsummer, where it is enjoyed with schnapps.

In the middle ages the Dutch developed a special treat known in English as soused herring.

Pickled herrings are also common in Ashkenazi Jewish cuisine, perhaps best known for *forshmak* salad known in English simply as "chopped herring".

Pickled herring can also be found in the cuisine of Hokkaido in Japan, where families traditionally preserved large quantities for winter.

Rollmops

The word Rollmops, borrowed from German, refers to a pickled herring fillet rolled (hence the name) into a cylindrical shape around a piece of pickled cucumber or an onion.

Fermented

In Sweden, Baltic herring is fermented to make surstromming.

Raw

A typical Dutch delicacy is raw herring (Hollandse Nieuwe). This is typically eaten with raw onions. Hollandse nieuwe is only available in Spring, when the first seasonal catch of herring is brought in.

This is celebrated in festivals such as the Vlaardingen Herring Festival. The new herring are frozen and enzyme-preserved for the remainder of the year.

Herring is also canned and exported by many countries. A *sild* is an immature herring that are canned as sardines in Norway or Denmark.

Very young herring are called whitebait and are eaten whole as a delicacy.

Other Means

A kipper is a split and smoked herring, a bloater is a whole smoked herring, and a buckling is a hot smoked herring with the guts removed. All are staples of British cuisine. According to George Orwell in *The Road to Wigan Pier*, the Emperor Charles V erected a statue to the inventor of bloaters.

In Scandinavia, Herring soup is also a traditional dish.

In Southeast Alaska, western hemlock boughs are cut and placed in the ocean before the herring arrive to spawn. The fertilized herring eggs stick to the boughs, and are easily collected.

After being boiled briefly the eggs are removed from the bough. Herring eggs collected in this way are eaten plain or in herring egg salad. This method of collection is part of Tlingit tradition.

Herring Lore

Figuratively, a *red herring* is a false lead in a mystery. In this context, *red* means smoked, and a smoked herring has such a strong smell that it can be used to create a false scent that causes hunting dogs to lose a track.

Herring Smelt

The herring smelts or argentines are a family, Argentinidae, of osmeriform fishes. They are similar in appearance to smelts (family Osmeridae), but have much smaller mouths.

They are found in oceans throughout the world. They are small fishes, growing up to 25 cm long, excepting the Greater argentine, *Argentina silus*, which reaches 70 cm.

They form large schools close to the sea floor and feed on plankton, especially krill, amphipods, small cephalopods, chaetognaths and ctenophores.

Several species are fished commercially and processed into fish meal.

Species

There are 25 species in two genera:

- Genus *Argentina*
  - Alice argentina, *Argentina aliceae* (Cohen & Atsaides, 1969).
  - *Argentina australiae* (Cohen, 1958).
  - Bruce's argentine, *Argentina brucei* (Cohen & Atsaides, 1969).
  - *Argentina elongata* (Hutton, 1879).
  - *Argentina euchus* (Cohen, 1961).
  - *Argentina georgei* (Cohen & Atsaides, 1969).
  - *Argentina kagoshimae* (Jordan & Snyder, 1902).
  - North-Pacific argentine, *Argentina sialis* (Gilbert, 1890).
  - Greater argentine, *Argentina silus* (Ascanius, 1775).
  - *Argentina sphyraena* (Linnaeus, 1758).
  - *Argentina stewarti* (Cohen & Atsaides, 1969).
  - Striated argentine, *Argentina striata* (Goode & Bean, 1896).
- Genus *Glossanodon*
  - *Glossanodon australis* (Kobyliansky, 1998).
  - *Glossanodon danieli* (Parin & Shcherbachev, 1982).

— *Glossanodon elongatus* (Hutton, 1879).
— Small-toothed argentine, *Glossanodon leioglossus* (Valenciennes, 1848).
— *Glossanodon lineatus* (Matsubara, 1943).
— *Glossanodon melanomanus* (Kobyliansky, 1998).
— *Glossanodon mildredae* (Cohen & Atsaides, 1969).
— *Glossanodon nazca* (Parin & Shcherbachev, 1982).
— *Glossanodon polli* (Cohen, 1958).
— *Glossanodon pseudolineatus* (Kobyliansky, 1998).
— Pygmy argentine, *Glossanodon pygmaeus* (Cohen, 1958).
— *Glossanodon semifasciatus* (Kishinouye, 1904).
— Struhsaker's deep-sea smelt, *Glossanodon struhsakeri* (Cohen, 1970).

Hillstream Loach

The hillstream loaches are a family Balitoridae of small Eurasian fish. Most species are rheophilic, living in swift, clear and well-oxygenated streams. Several species of the subfamily Balitorinae live in fast-flowing streams or torrents and have a large sucker mouth and modified ventral fins used for clinging to rocks. They have a number of similarities with the sibling family of loaches (Cobitidae), such as multiple barbels around the mouth. They should not be confused with the loricariids, which looks similar but is a family of catfish. The family includes over 600 species in more than 60 genera. Many of the species are popular for aquaria.

Classification

- Subfamily Balitorinae
  — *Annamia*
  — *Balitora*
  — *Balitoropsis*
  — *Beaufortia*
  — *Bhavania*
  — *Crossostoma*

— *Cryptotora*
— *Dienbiena*
— *Erromyzon*
— *Gastromyzon*
— *Glaniopsis*
— *Hemimyzon*
— *Homaloptera*
— *Homalosoma*
— *Hypergastromyzon*
— *Jinshaia*
— *Katibasia*
— *Lepturichthys*
— *Liniparhomaloptera*
— *Metahomaloptera*
— *Neogastomyzon*
— *Neohomaloptera*
— *Paraprotomyzon*
— *Parhomaloptera*
— *Plesiomyzon*
— *Protomyzon*
— *Pseudogastromyzon*
— *Pseudohomaloptera*
— *Sectoria*
— *Sewellia*
— *Sinogastromyzon*
— *Sinohomaloptera*
— *Travancoria*
— *Vanmanenia*

- Subfamily Nemacheilinae

— *Aborichthys*
— *Acanthocobitis*
— *Adiposia*
— *Barbatula*

— *Barbucca*
— *Dzihunia*
— *Ellopostoma*
— *Eonemachilus*
— *Heminoemacheilus*
— *Ilamnemacheilus*
— *Indoreonectes*
— *Lefua*
— *Longischistura*
— *Mesonoemacheilus*
— *Micronemacheilus*
— *Nemacheilus*
— *Nemachilichthys*
— *Neonoemacheilus*
— *Nun*
— *Oreonectes*
— *Orthrias*
— *Oreias*
— *Paracobitis*
— *Paranemachilus*
— *Physoschistura*
— *Protonemacheilus*
— *Schistura*
— *Seminemacheilus*
— *Sphaerophysa*
— *Sundoreonectes*
— *Traccatichthys*
— *Triplophysa*
— *Tuberoschistura*
— *Turcinoemacheilus*
— *Vaillantella*
— *Yunnanilus*

Horn Shark

The horn shark, *Heterodontus francisci,* is a bullhead shark. It can reach a size of 122 cm, and is brown with black spots. Its range is from central California to the Gulf of California, Mexico, and probably also in Ecuador and Peru. It is mostly nocturnal, and appears sluggish in the daytime.

Its habitat includes rocky reefs, kelp beds, sand flats, crevices, and caverns in a depth range from 2 to 150 m. Adults tend to return to the same resting spot every day. It feeds on invertebrates, primarily sea urchins, crabs, probably abalone, and other fish.

The horn shark is oviparous, laying 12-14 cm (5 inches) long kelp-coloured spiral-shaped eggs that may float up on beach shores.

Hound Sharks

Hound sharks are a family, Triakidae, of ground sharks, consisting of about 40 species in 9 genera. In some classifications, the family is split in two, with *Mustelus, Scylliogaleus,* and *Triakis* in Triakidae, and the remaining genera in Galeorhinidae.

Hound sharks are small sharks with two large spineless dorsal fins and an anal fin. They are found throughout the world in warm and temperate waters, where they feed on fishes and invertebrates on the seabed and in midwater.

Genera

- *Furgaleus*
- *Galeorhinus*
- *Gogolia*
- *Hemitriakis*
- *Hypogaleus*
- *Iago*
- *Mustelus*
- *Scylliogaleus*
- *Triakis*

Cladogram

- Triakidae
  - — Triakinae
    - ⇒ Scylliogaleini
    - ⇒ Scylliogaleus
    - ⇒ Triakini
    - ⇒ Mustelus
    - ⇒ Triakis
    - ⇒ Subgenus Triakis
    - ⇒ Subgenus Cazon (may be a genus)
  - — Galeorhininae
    - ⇒ Galeorhinini
    - ⇒ Galeorhinus
    - ⇒ Hypogaleus
    - ⇒ Iagini
    - ⇒ Gogolia
    - ⇒ Hemitriakis
    - ⇒ Iago

Huchen

The huchen or Danube salmon (*Hucho hucho*) is a species of freshwater fish in the salmon family (family Salmonidae) of order Salmoniformes. It is the type species of its genus. Native to the Palearctic ecozone, the huchen occurred originally in the Danube basin in Europe but has been introduced elsewhere on the continent. This food fish is threatened with extinction.

Appearance

The huchen has a slender body that is nearly round in cross-section. On the reddish brown back are several dark patches in an X or crescent shape. Smaller fish feed on the larvae of water insects or on insects dropped into the water; the larger individuals are predators of other species of fish and other small vertebrates such as mice crossing rivers.

Reproduction

This largest permanent fresh water salmonid spawns in April, once water reaches a temperature of 6 to 9 °C. For spawning, the huchen migrates up the river, where females excavate depressions in the gravel in which to deposit the eggs. Larvae hatch 30 to 35 days after fertilization.

Commercial Breeding

There is now a considerable effort to commercially produce huchen larvae for reintroduction into the wild. This requires the adults being caught just before spawning and kept in special tanks. Fry are released in appropriate places once they have reached 4 to 10 cm.

Reef Triggerfish

The reef, rectangular, or wedge-tail triggerfish is one of several species of triggerfish.

Classified as *Rhinecanthus rectangulus,* it is endemic to the salt water coasts of various central and south Pacific Ocean islands. It is often asserted that the Hawaiian name is one of the longest words in the English Language and that "the name is longer than the fish."

Description

The triggerfish's teeth are blue set close together inside its relatively fat mouth, and it has a small second spine, which it can use to lock its first spine into an upright position. The triggerfish will wedge itself into small crevices and lock its spine to make it almost impossible to get out. In addition, when fleeing predators, the triggerfish will sometimes make grunting noises, possibly a mating call to warn other nearby triggerfish of danger at hand. They hide in crevasses.

Distribution

The reef triggerfish is distributed throughout the Indo-Pacific region, and it is especially prominent in the coral reefs of the Hawaiian Islands.

Hawaii State Fish

Due to an expiration of a Hawaiian state law, the trigger fish ceased to be the state fish of Hawaii in 1990. However, as of April 2006, a bill was presented to the Governor of Hawaii which reinstated the reef triggerfish as the state fish of Hawaii. The bill passed into law May 2, 2006 and was effective upon its approval.

Humuhumunukunukuapua a means "triggerfish with a pig-like short snout". It is not, as often claimed, the longest fish name in Hawaiian; that distinction belongs to *lauwiliwilinukunuku* ("long-snouted fish shaped like a *wiliwili* leaf"), the butterflyfish *Forcipiger longirostris.*

Ide

The ide or orfe, *Leuciscus idus*, is a freshwater fish of the family Cyprinidae found across northern Europe and Asia. It occurs in larger rivers, ponds, and lakes, typically in schools.

The body has a typical cyprinid shape and generally silvery appearance, while all the fins are red in varying degrees, particular in the "golden ide" or "golden orfe" variety.

Ides are predators, eating insects, crustaceans, molluscs, and small fish. In the spring, they move into rivers to spawn over gravel or vegetation; the eggs may be found sticking to stones or weeds in shallow water.

Spread

Orfe eggs, derived from ornamental pond stocks, were illegally imported to New Zealand by mail sometime in the 1980s. Subsequent releases occurred between 1985–86 in at least 8 and possibly 5 more sites north of Auckland.

The current status of these populations is in doubt, and at least one release site remains unknown. It seems likely to persist in the wild in New Zealand. Whether they become a nuisance species in New Zealand or will be successfully eradicated remains to be seen.

Inconnu

The inconnu or sheefish (*Stenodus leucichthys*) is a large salmonid fish, related to the whitefishes, found in North America, Europe and the Middle East.

The fish has a large mouth with a protruding lower jaw and a high and pointed dorsal fin. It is generally silver in colour with green, blue or brown on the back. The meat is white, flaky and somewhat oily.

The fish eat plankton for their first year of life and then become predators of smaller fish. The fish live in lakes and rivers and in the brackish water at the outlets of rivers into the ocean. The fish may migrate 1500 km (1000 miles) to spawn.

An adult fish weighs from 14-25 kg (30-55 pounds). IUCN lists the inconnu as Endangered.

Jackfish

Jackfish may refer to:

- The northern pike (*Esox lucius*) in Canada.
- The northern pike or the banded rudderfish (*Seriola zonata*) in the United States.
- The trevally (genus *Caranx*) in New Zealand.
- The bigeye scad (*Selar crumenopthalmus*) or crevalle jack (*Caranx hippos*) in the Caribbean.
- The ghost town of Jackfish, Ontario in Canada.

There are also several Jackfish Rivers.

Jack Dempsey

The Jack Dempsey (*Archocentrus octofasciatus* (Regan, 1903), formerly *Nandopsis octofasciatum/Cichlasoma octofasciatum*) is a cichlid fish named for the 1920s boxer Jack Dempsey. The name alludes to its aggressive nature. Like most cichlids, it is territorial, especially against its own kind and similar species. The fish was once very popular due to its striking appearance and personable mannerisms. While it is a popular aquarium fish, due to its behaviour and size it is not easy to keep.

The colouration changes as the fish matures from a light gray or tan with faint turquoise flecks to a dark purple-gray with very bright, iridescent blue, green, and gold flecks. The dorsal and anal fins of mature males have long, pointed tips. Females lack these exaggerated tips.

The fish is native to Yucatan and Central America, where it is found in slow-moving waters, such as swampy areas with warm, murky water, weedy, mud- and sand-bottomed canals, drainage ditches, and rivers. It is also established as an introduced species in Australia, the USA and Thailand (presumably as an aquarium escapee). The Jack Dempsey natively lives in a tropical climate and prefers water with a 7–0 pH, a water hardness of 9–20 dGH, and a temperature range of 72–86°F (22–30°C). It can reach up to 25 cm (10 in) in length. It is carnivorous, eating worms, crustaceans, insects and other fish.

Jack Dempseys lay their eggs on the substrate (the bottom of the aquarium or pool). Like most cichlids, they show substantial parental care: both parents help incubate the eggs and guard the fry when they hatch. Jack Dempseys are known to be attentive parents, pre-chewing food to feed to their offspring. A lot of times however, they will eat their fry when the breeding pair are overly disturbed or something in their environment is wrong.

In 1997, the *San Francisco Chronicle* reported that a man had died when he put a Jack Dempsey into his mouth as a joke: the fish presumably erected its fin spines to avoid being swallowed, a characteristic cichlid anti-predator response, and became wedged in the man's throat.

Japanese eel

The Japanese eel, *Anguilla japonica,* is a species of eel found in Japan, Korea, the East China Sea and the northern Philippines. Like the European eel it lives in both fresh water and the sea. The spawning grounds are presumed to be the western slopes of the Suruga seamount. They are eaten in Japan as unagi, and also have uses in Chinese medicine.

4

Jawfish

Opistognathidae

Opistognathidae (*opisto* = "behind", *gnath* = "mouth"), commonly referred to as jawfishes, are classified within Order Perciformes, Suborder Percoidei. They are found throughout shallow reef areas of the Atlantic and Pacific Oceans.

Physically similar to blennies, jawfish are generally smaller-sized fish with an elongated body plan. Their heads, mouths, and eyes are large in size relative to the rest of their bodies. Jawfish possess a single, long dorsal fin with 9-12 spines and a caudal fin that can be either rounded or pointed.

Jawfish typically reside in burrows that they construct in sandy substrate. They will stuff their mouth with sand and spit it out elsewhere, slowly creating a tunnel. Utilising the protection of these burrows, these fish will hover feeding on plankton or other small organisms, ready to dart back in at the first sign of danger. They are territorial of the area around their burrows. Jawfish are mouthbrooders meaning that their eggs hatch in their mouths, where the new-born fry are able to be protected from predators.

Species

As of 2006, there are sixty described species in four genera. There are thought to be many undescribed species.

- Genus *Lonchopisthus*
  - — *Lonchopisthus higmani* (Mead, 1959).
  - — *Lonchopisthus lemur* (Myers, 1935).
  - — *Lonchopisthus lindneri* (Ginsburg, 1942).
  - — Swordtail jawfish, *Lonchopisthus micrognathus* (Poey, 1860).
  - — Longtailed jawfish, *Lonchopisthus sinuscalifornicus* (Castro-Aguirre & Villavicencio-Garayzar, 1988).
- Genus *Merogymnoides*
  - — *Merogymnoides carpentariae* (Whitley, 1966).
- Genus *Opistognathus*
  - — Yellowhead jawfish, *Opistognathus aurifrons* (Jordan & Thompson, 1905).
  - — Darkfin jawfish, *Opistognathus brasiliensis* (Smith-Vaniz, 1997).
  - — *Opistognathus castelnaui* (Bleeker, 1860).
  - — Bartail jawfish, *Opistognathus cuvierii* (Valenciennes, 1836).
  - — Darwin jawfish, *Opistognathus darwiniensis* (Macleay, 1878).
  - — *Opistognathus decorus* (Smith-Vaniz & Yoshino, 1985).
  - — *Opistognathus dendriticus* (Jordan & Richardson, 1908).
  - — *Opistognathus evermanni* (Jordan & Snyder, 1902).
  - — *Opistognathus eximius* (Ogilby, 1908).
  - — *Opistognathus fenmutis* (Acero P. & Franke, 1993).
  - — Galapagos jawfish, *Opistognathus galapagensis* (Allen & Robertson, 1991).
  - — Yellow jawfish, *Opistognathus gilberti* (Bohlke, 1967).
  - — *Opistognathus hongkongiensis* (Chan, 1968).
  - — *Opistognathus hopkinsi* (Jordan & Snyder, 1902).

— Black jawfish, *Opistognathus inornatus* (Ramsay & Ogilby, 1887).
— *Opistognathus iyonis* (Jordan & Thompson, 1913).
— Smiler, *Opistognathus jacksoniensis* (Macleay, 1881).
— Blotched jawfish, *Opistognathus latitabundus* (Whitley, 1937).
— Roughcheek jawfish, *Opistognathus leprocarus* (Smith-Vaniz, 1997).
— *Opistognathus liturus* (Smith-Vaniz & Yoshino, 1985).
— Moustache jawfish, *Opistognathus lonchurus* (Jordan & Gilbert, 1882).
— Banded jawfish, *Opistognathus macrognathus* (Poey, 1860).
— *Opistognathus macrolepis* (Peters, 1866).
— Halfscaled jawfish, *Opistognathus margaretae* (Smith-Vaniz, 1984).
— Mottled jawfish, *Opistognathus maxillosus* (Poey, 1860).
— Largesacle jawfish, *Opistognathus megalepis* (Smith-Vaniz, 1972).
— Yellowmouth jawfish, *Opistognathus melachasme* (Smith-Vaniz, 1972).
— Mexican jawfish, *Opistognathus mexicanus* (Allen & Robertson, 1991).
— Robust jawfish, *Opistognathus muscatensis* (Boulenger, 1887).
— Birdled jawfish, *Opistognathus nigromarginatus* (Ruppell, 1830).
— *Opistognathus nothus* (Smith-Vaniz, 1997).
— Panamanian jawfish, *Opistognathus panamaensis* (Allen & R Robertson, 1991).
— Papuan jawfish, *Opistognathus papuensis* (Bleeker, 1868).
— Finespotted jawfish, *Opistognathus punctatus* (Peters, 1869).

— Leopard jawfish, *Opistognathus reticulatus* (McKay, 1969).
— Giant jawfish, *Opistognathus rhomaleus* (Jordan & Gilbert, 1882).
— Spotfin jawfish, *Opistognathus robinsi* (Smith-Vaniz, 1997).
— *Opistognathus rosenbergii* (Bleeker, 1857).
— Blue-spotted jawfish, *Opistognathus rosenblatti* (Allen & Robertson, 1991).
— Bullseye jawfish, *Opistognathus scops* (Jenkins & Evermann, 1889).
— *Opistognathus signatus* (Smith-Vaniz, 1997).
— *Opistognathus solorensis* (Bleeker, 1853).
— Dusky jawfish, *Opistognathus whitehursti* (Longley, 1927).

- Genus *Stalix*

— *Stalix davidsheni* (Klausewitz, 1985).
— *Stalix dicra* (Smith-Vaniz, 1989).
— *Stalix eremia* (Smith-Vaniz, 1989).
— *Stalix flavida* (Smith-Vaniz, 1989).
— *Stalix histrio* (Jordan & Snyder, 1902).
— *Stalix immaculata* (Xu & Zhan, 1980).
— *Stalix moenensis* (Popta, 1922).
— Oman jawfish, *Stalix omanensis* (Norman, 1939).
— *Stalix sheni* (Smith-Vaniz, 1989).
— *Stalix toyoshio* (Shinohara, 1999).
— *Stalix versluysi* (Weber, 1913).

Jellynose Fish

The jellynose fishes are a small order (Ateleopodiformes) of ray-finned fish, consisting of a single family (Ateleopodidae) with about a dozen species in four genera.

Jellynoses are deep-water marine fish. Their skeletons are largely cartilage (thus the name), although they are true teleosts, and not at all related to Chondrichthyes. Heads are large, with a bulbous nose, and the (usually) elongated body tapers towards the tail. Their caudal fins are very small, and merged with long anal fins, and the pelvic fins are single rays, except for *Guentherus*. Dorsal fins tend to be prominent and placed just behind the head. The species have a range of sizes, the longest reaching 2 metres.

They are found in the Caribbean Sea, eastern Atlantic Ocean, and the Indo-Pacific area. Most of the species are poorly-known, but the highfin tadpole fish, *Guentherus altivelis*, is of potential interest for commercial fishing.

Species

- Genus *Ateleopus*
  - — *Ateleopus indicus* (Alcock, 1891).
  - — *Ateleopus japonicus* (Bleeker, 1854).
  - — *Ateleopus natalensis* (Regan, 1921).
  - — *Ateleopus purpureus* (Tanaka, 1915).
  - — *Ateleopus tanabensis* (Tanaka, 1918).
- Genus *Guentherus*
  - — Highfin tadpole fish, *Guentherus altivela* (Osorio, 1917).
- Genus *Ijimaia*
  - — *Ijimaia antillarum* (Howell Rivero, 1935).
  - — *Ijimaia dofleini* (Sauter, 1905).
  - — *Ijimaia fowleri* (Howell Rivero, 1935).
  - — Loppe's tadpole fish, *Ijimaia loppei* (Roule, 1922).
  - — Deep water ateleopid fish, *Ijimaia plicatellus* (Gilbert, 1905).
- Genus *Parateleopus*
  - — *Parateleopus microstomus* (Smith & Radcliffe, 1912).

Jewel Cichlid

Hemichromis is a genus of fish from the family Cichlidae, known in the aquarium trade as Jewel cichlids. Jewel cichlids are native to west Africa. Within west Africa *Hemichromis* species are found in creeks, streams, rivers and lakes with a variety of water qualities.

Jewel cichlids can reach lengths of 7-13 cm (6inch) in some species brightly coloured. Brighter body colouration is generally evident during breeding.

Sexual dimorphism is limited, though male jewel cichlids are typically more brightly coloured and in some species have more pointed anal, ventral and dorsal fins.

Like most cichlids, jewel cichlids have high developed brood care. *Hemichromis* species typically form monogamous breeding pairs and the female spawns on a flat surface such as a leaf or stone. Both parents guard the eggs, and participate in fry raising.

Species

- *Hemichromis angolensis* (Steindachner, 1865)
- *Hemichromis bimaculatus* (Gill, 1862)
- *Hemichromis cerasogaster* (Boulenger, 1899)
- *Hemichromis elongatus* (Guichenot, 1861)
- *Hemichromis exsul* (Trewavas, 1933)
- *Hemichromis fasciatus* (Peters, 1857)
- *Hemichromis frempongi* (Loiselle, 1979)
- *Hemichromis letourneuxi* (Sauvage, 1880)
- *Hemichromis lifalili* (Loiselle, 1979)
- *Hemichromis stellifer* (Loiselle, 1979)

Aquarium Care

Jewel cichlids are neither suited to beginners, nor the usual community tank. Several young specimens may be kept in a spacious aquarium, with stones and wood for cover until a pair forms prior to breeding. Their innate aggression makes them good candidates for keeping in a monospecies aquarium. Jewel cichlids are omnivorous and will eat both live foods and fish flakes.

Jewfish

Jewfish is a term broadly applied to a number of species of marine fish, including:

Family	*Species*
Glaucosomatidae (Perciformes):	Weststralian dhufish *Glaucosoma hebraicum.*
Megalopidae (Elopiformes):	The tarpon (*Megalops atlanticus*).
Paralichthyidae (Pleuronectiformes):	The summer flounder (*Paralichthys dentatus*).
Polyprionidae (Perciformes):	The giant sea bass (*Stereolepis gigas*); The wreckfish (*Polyprion americanus*).
Sciaenidae (Perciformes):	The Japanese meagre (*Argyrosomus japonicus*); Black Jewfish (Protonibea diacanthus); Serranidae (Perciformes). The Warsaw grouper (*Epinephelus nigritus*), sometimes called the black jewfish; the Goliath grouper (*Epinephelus itajara*).

It was also sometimes loosely applied to various other groupers of family Serranidae.

The term red Jew could refer to either of two lutjanids (Perciformes), the crimson snapper (*Lutjanus erythropterus*) or the Malabar blood snapper (*Lutjanus malabaricus*).

In 2001 the American Fisheries Society renamed the jewfish (*Epinephelus itajara*), the goliath grouper out of concern for the potential offensiveness of the name.

Australian Salmon

Australian salmon or Australasian salmon, also known as kahawai in New Zealand English, are medium-sized perciform marine fish of the small family Arripidae (also spelled Arripididae). Four species are recognised, all within the genus

Arripis; they are found in the waters off southern Australia (including Tasmania) and New Zealand. Despite the common name, Australian salmon are not related to the salmon (family Salmonidae) of the Northern Hemisphere; the former were named so by early European settlers after their superficial resemblance to the salmoniform fishes.

Relatively long-lived fish, Australian salmon are a favoured target of recreational fishers, and both commercial and traditional Maori fisheries; the fish are also common bycatch of the snapper, mullet, trevally, and mackerel fisheries. These species are all taken in great numbers by way of purse seine nets and trawling. They are also caught by artisanal fishermen along the southern coastline of Australia by beach seining. Due to declining numbers and ever-increasing annual catch sizes, the future viability of the Australian salmon stock has been put into question.

Species and Range

The ranges of the four species may overlap to some extent, but can be described thus:

- *Australian herring, Australian ruff or tommy ruff, Arripis georgianus:* Gippsland Lakes, Victoria to Shark Bay, Western Australia; endemic.
- *Eastern Australian salmon, Arripis trutta:* From western Victoria to New Zealand, including the islands of Tasmania, Lord Howe, and Norfolk (rarely to Western Australia).
- *Western Australian salmon, Arripis truttacea:* Western Australia to Victoria and Tasmania.
- *Northern kahawai, Arripis xylabion:* New Zealand, west to Lord Howe, Norfolk and Kermadec Islands.

It should be noted that some systematists consider *A. trutta* and *A. truttacea* as subspecies of the same species.

Physical Description

A streamlined, fusiform body and large, powerful forked tail—the upper lobe of which is equal to or less than the length of the head in the Eastern Australian salmon, *Arripis trutta*—are indications of the fast-paced pelagic lives these fish lead. The first (spinous, with nine spines) dorsal fin originates behind the pectoral fins, the former being confluent with, but noticeably higher than the much longer soft dorsal fin (with 15 - 19 rays), itself much longer than the anal fin (which has three spines and 9 - 10 soft rays). The pelvic fin is situated in a thoracic position.

Scales and eyes are relatively large—in the Australian herring, *Arripis georgianus*, the eyes are approximately one fifth the length of the head in diameter. The large mouth is terminal, and the jaws are lined with bands of sharp villiform (brushlike) teeth. The lateral line runs along the upper sides of the body.

The Western Australian salmon, *Arripis truttacea*, is the largest species at a maximum 96 cm (three feet) standard length (that is, excluding the caudal fin) and 10.5 kg in weight. The Australian herring is the smallest species at maximum 41 cm (16 inches) fork length (that is, from the snout to the middle of the caudal fin's fork) and 800 g. Australian salmon share a passing resemblance to the unrelated yellowtail amberjack, *Seriola lalandi*, locally known as "kingfish", with which larger salmon are sometimes confused.

All species are strongly countershaded; dorsal colours range from dark blue-green in *A. trutta*, green in *A. georgianus*, and steel-blue to greyish- or greenish-black in *A. truttacea*; the colours fade to a silver-white ventrally. A smattering of yellow, gray, or blackish spots embellishes the dorsal half, the spots arranged vertically or longitudinally in a series of rows. There are marked differences in subadult colouration: for example, on the flanks of juvenile Australian herring are a series of dark golden vertical bars.

Reproduction

Although their reproduction is poorly studied, Australian salmon are all known to be "pelagic spawners"; that is, they spawn in open water during the austral spring, releasing a large number of tiny (1 mm in diameter), smooth spherical eggs made buoyant by lipid droplets. The eggs (and later the larvae), which possess an unsegmented yolk, become part of the zooplankton, drifting at the mercy of the currents until the larvae develop and settle.

The young salmon then spend the first 2 - 5 years in sheltered coastal bays, inlets, and estuaries until they become sexually mature and begin to move into more open waters. Relatively long-lived fishes, Australian salmon may attain an age of 26 years in *Arripis trutta* and 7 - 9 years in other species.

All species are oceanic spawners. Reports of *A. trutta* being anadromous and spawning in freshwater are not correct; this may be due to confusion with sea run specimens of exotic brown trout, *S. trutta*, or anadromous populations of native spotted mountain trout, *Galaxias truttaceus*.

Arripis georgianus are thought—due to females retaining both ripe and unripe eggs—to be "partial spawners"; that is, they may spawn over a long period with no real peaks. In contrast, *A. trutta* and *A. truttacea* are thought to be "serial batch spawners", completing their spawning season after a series of small "burst" spawnings.

Ecology

All species are neritic and epipelagic, staying within the upper layers of relatively shallow (from 1 - 80 metres), open and clear coastal waters (although the Western Australian salmon may prefer deeper water). The turbulent surf zone of beaches, rocky reefs, bays, and brackish waters such as estuaries are also frequented, and some species may also enter rivers. Juveniles inhabit estuaries and mangroves, as well as sheltered bays with soft bottoms carpeted with seagrass. Adults undertake seasonal migrations over vast distances, moving into deeper water during winter months.

Australian salmon form immense schools with hundreds to thousands of individuals, as both adults and juveniles. They are carnivorous and feed primarily on small fish such as pilchard; crustaceans such as krill, copepods, and other zooplankton (the latter comprising the bulk of the juvenile diet). The zoobenthos is also sampled to some extent, with primarily shellfish, crabs, and annelid worms eaten. The salmon are very fast swimmers, and are sometimes seen mingling with ostensibly similar species of carangids, such as trevally; this is an example of mutualism.

Together with the carangids, Australian salmon feed en masse by cooperatively bullying baitfish up to the surface; this herding technique is exploited by seabirds which are quickly attracted to, and feed upon, the foaming mass of fish at the surface. This commensal relationship between the salmon and the birds is noted to be especially strong in such species as the White-fronted Tern, *Sterna striata,* Fluttering Shearwater, *Puffinus gavia,* and Buller's Shearwater, *Puffinus bulleri.* The baitfish made available by the salmon's herding behaviour may also be important to the reproductive success of winter-nesting birds; the decline of the salmon stocks has evoked concern for these bird species, some of which - such as the Fairy Tern, *Sterna nereis,* - are endangered.

Aside from seabirds, the salmon are also important in the diets of cetaceans such as Orca and Bottlenose Dolphins; several species of large sharks, for example; great white, dusky, copper, and sand tiger sharks; and eared seals such as the Australian Sea Lion.

Importance to Humans

Pungently flavoured, coarse, and slightly oily flesh makes Australian salmon less desirable as a food fish; it is often sold canned or is smoked to improve its flavour, and bleeding the fish out is also said to help.

What is not sold for human consumption is used as bait for rock lobster traps and other commercial and recreational fishing. The salmon fetch no more than a few dollars (AU) per kilogram; nonetheless, large numbers are taken via purse seine nets (and to a lesser extent trawling, hauling, gill, and trap nets) annually; the reported 2002 - 03 commercial New Zealand catch of kahawai was 2,900 tonnes. Such reported catches do not include the untold tonnes taken as bycatch from operations targeting more highly valued species.

Low-flying planes are used to locate and target sizeable salmon schools, and critics have cited this practice as a means by the industry to artificially inflate catch records (which would give a false impression of abundance). Salmon numbers have declined noticeably however, with large specimens becoming ever rarer; the fish have all but disappeared from some areas. It was not until October 1st, 2004 that the New Zealand Ministry of Fisheries included kahawai under its Quota Management System, setting a catch limit of 3,035 tonnes for the season. This was a five per cent increase over the previous two years, despite the government's intention of lowering catch limits.

The native Maori of New Zealand fish for the salmon in a subsistence and customary capacity, to whom the fish are known as *kahawai, koopuuhuri,* and *kooukauka*. The fish were (and are) caught with lines of flax fibre and elaborate hooks of bone, wood, shell such as abalone (paua), or stone. The salmon are filleted before being hung on racks to dry. Recreational fishers also seek Australian salmon for their renowned mettle when hooked; the salmon are a challenge to land and often jump, occasionally standing on their tails. A significant number are taken for sport. No records of total recreational catches are kept, but the year's estimated catch of Australian herring from Western Australia's Blackwood River estuary beginning May, 1974 was 68,000 individuals.

Kaluga

The Kaluga (*Huso dauricus*) is a large predatory sturgeon found in the Amur River basin. Also known as the Great Siberian Sturgeon, they are claimed to be the largest freshwater fish in the world, with a maximum weight of at least 1000 kg. *Kaluga caviar* comes from the Kaluga sturgeon.

Kelpfish

The kelpfishes are a family of perciform fishes, native to coastal Australia and New Zealand.

The name is from Greek *cheir* meaning "hands" and *nema* meaning "thread".

There are six species in two genera:

- Genus *Chironemus*
  - — *Chironemus bicornis* (Steindachner, 1898).
  - — *Chironemus delfini* (Porter, 1914).
  - — Tasselled kelpfish, *Chironemus georgianus* (Cuvier, 1829).
  - — Large kelpfish, *Chironemus marmoratus* (Gunther, 1860).
  - — *Chironemus microlepis* (Waite, 1916).
- Genus *Threpterius*
  - — Silver spot, *Threpterius maculosus* (Richardson, 1850).

Killifish

A killifish is any of various small, mainly oviparous (egg-laying) cyprinodont fish (order Cyprinodontiformes, family Cyprinodontidae). The name *killifish* is derived from the Dutch word "kilde", meaning *small creek, puddle*. Most killies are small fish, one to two inches (2.5 to 5 cm). The largest is under six inches (15 cm), but only a few species are this large. Although *killifish* is sometimes used as an English equivalent to Cyprinodontidae, some species belonging to this family have their own common names, such as the pupfish and the mummichog.

Range and Habitat

Killifish are found mainly in fresh or brackish waters in the Americas, as far south as Argentina and as far north as southern Ontario. There are also species in southern Europe, in much of Africa as far south as Kwa-Zulu Natal, South Africa, in the Middle East and Asia (as far east as Vietnam), and on several Indian Ocean islands. Killifish are not found in Australia, Antarctica, or northern Europe.

The majority of killifish are found in permanent streams, rivers, and lakes, and live between two and three years. Such killifish are common in the Americas (*Cyprinodon, Fundulus* and *Rivulus*) as well as in Africa and Asia (*Aphyosemion,*

Aplocheilus, Epiplatys, Fundulopanchax, Lacustricola, etc.) and southern Europe (*Aphanius*).

Some specialised forms live in temporary ponds and flood plains, and typically have a much shorter lifespan. Such "annual" species live no longer than nine months, and are used as models for studies on ageing.

Examples include the African genus *Nothobranchius* and South American genera ranging from the cold water *Austrolebias* of Argentina and Uruguay to the more tropical *Gnatholebias, Pterolebias, Simpsonichthys* and *Terranatos*.

Territorial Behaviour

A small number of species will shoal while most are territorial to varying degrees. Population can be dense and territories can shift quickly, especially for species of the extreme shallows (a few centimetres of water).

Many species exist as passive tribes in small streams where dominant males will defend a territory while allowing females and immature males to pass through the area. In the aquarium, territorial behaviour is different for every grouping, and will even vary by individuals. In a large enough aquarium, most species can live in groups as long as there are more than three males.

Diet

Killifish feed primarily on aquatic arthropods such as insect (mosquito) larvae, aquatic crustaceans and worms. It is reported by the killifish collector Rudolf Koubek that areas in Gabon where the streams lack killifish (due to pollution or other causes) are rife with malaria, which is spread by a mosquito. Some species of *Orestias* from Lake Titicaca are planktonic filter feeders. Others, such as *Cynolebias* and *Megalebias* species and *Nothobranchius ocellatus* are predatory and feed mainly on other fish.

Reproduction

Reproductive strategies of killifish are diverse. Some will spawn in no specific location while a small number of them (the *Lampeyes*) will spawn at specific sites or on specific environmental structures (e.g. *Lamprichthys tanganicanus* in rock crevices). Some species (e.g. *Cyprinodon*) will establish "nests" on the substrate wherein the male and various females will spawn. The annual killifish will spawn in the mud of the temporary ponds and floodplains.

In the mud the eggs will cease development (diapause) dependent on the environmental conditions. There is some evidence that where the ponds do not dry out the eggs will develop as normal and hatch in the water (semi-annual). Where the ponds dry out the eggs will lie semi-dormant, developing slowly until the ponds fill again whereupon the fry will hatch from the eggs and establish a new generation.

Males of some species in the genera *Campellolebias* and *Cynopoecilus* possess a gonopodium and practice a form of pseudo-internal fertilization, and are thought to be an evolutionary pathway to the live bearing tooth carps (*livebearers*). The species *Rivulus marmoratus* is the only known naturally occurring, truly self-fertilizing vertebrate.

Furthermore, they are always either male or hermaphroditic; females of the species don't seem to exist. Only about 5 per cent of a population are born as males; after 3-4 years about 60 per cent of the (self-fertilizing) hermaphrodites transform into secondary males by losing female structures and functions. The proportion of males depends on the environmental temperature: below 20°C (68°F) the majority are males, and above 25°C (77°F) all specimens are hermaphrodites.

Killifish as Pets

Many killifish are lavishly coloured; and most species are easy to keep and breed in an aquarium. Specimens can be obtained from specialist societies such as the American Killifish Association, British Killifish Association or Deutsche Killifisch Gemeinschaft.

Killies are seldom if ever found in pet shops but are very common on the online fish auction site Aquabid.

King-of-the-Salmon

King-of-the-salmon, *Trachipterus altivelis*, is a ribbonfish of the family Trachipteridae. Found off the Pacific shore of North and South America from Alaska to Chile, generally at a depth of over 1,500 feet (460 m), it has a long tapering grey body with red fins and large eyes. The anal fin is absent. It grows to 183 cm.

Kissing Gourami

Kissing gouramis, also known as kissers (*Helostoma temminckii*), are large tropical freshwater fish comprising the monotypic gourami family Helostomatidae (from the Greek *elos* [stud, nail], *stoma* [mouth]). These fish originate from Thailand to Indonesia. They are highly commercial food fish which are farmed in their native South East Asia. They are used fresh for steaming, baking, broiling, and pan frying.

Physical Description

Typical of gouramis, the body is deep and strongly compressed laterally. The long-based dorsal (16–18 spinous rays, 13–16 soft) and anal fins (13–15 spinous rays, 17–19 soft) mirror each other in length and frame the body. The posterior most soft rays of each of these fins are slightly elongated to create a trailing margin. The foremost rays of the jugular pelvic fins are also slightly elongated. The pectoral fins are large, rounded, and low-slung. The caudal fin is rounded to concave. The lateral line is divided in two, with the posterior portion starting below the end of the other; there are a total of 43–48 scales running the line's length.

The most distinctive feature of the kissing gourami is its mouth. Other than being terminal rather than superior (as in other gourami families), the kissing gourami's mouth is highly protrusible; as its family name suggests, the lips are lined with horny teeth. However, teeth are absent from the premaxilla, dentaries, palatine, and pharynx. The gill rakers are also well-developed and numerous. The visible scales of the body are ctenoid, whereas the scales of the top of the head are cycloid.

Kissing gouramis reach a maximum size of 30 centimetres (12 in) TL. There is no outward sexual dimorphism and is difficult to almost impossible to distinguish the sexes.

There are two colour morphs encountered: greens, which have lengthwise lateral stripes and opaque, dark brown fins; and pink, which have a rose to orangy pink body and silvery scales, with transparent pinkish fins. Green fish originate from Thailand while pink fish origiante from Java. There is also a "dwarf" or "balloon pink" variety, which is a mutated strain of the pink gouramis that are offered to hobbyists. The "balloons" are named so for their smaller and rounder bodies.

Habitat and Ecology

Shallow, slow-moving, and thickly vegetated backwaters are the kissing gourami's natural habitat. They are midwater omnivores that primarily graze on benthic algae and aquatic plants, with insects taken from the surface. It is also a filter feeder, using their many gill rakers to supplement their diet with plankton. The fish use their toothed lips to rasp algae from stones and other surfaces. This rasping action, which (to humans) looks superficially like kissing, is also used by males to challenge the dominancy of conspecifics.

Spawning occurs from May to October in Thailand. Kissing gouramis are open-water egg scatterers; spawning is initiated by the female and takes place under cover of floating vegetation. The eggs, which the adults do not guard, are spherical, smooth, and buoyant. Initial development is rapid: the eggs hatch after one day, and the fry are free-swimming two days thereafter. The Kissing gourami is the only member of the suborder Anabantoidei not to build a nest or otherwise care for its young.

In the Aquarium

Kissing gouramis are also popular with aquarists for the fish's peculiar "kissing" behaviour of other fish, plants, and other objects. Large quantities of these fish are exported to Japan, Europe, North America, Australia, and other parts of the world for just this reason. Kissing gouramis need a roomy tank in order to thrive; they grow rapidly, and juvenile fish will quickly outgrow smaller aquaria. Kissing gouramis are tolerant towards fish of similar size. Male kissers will occasionally challenge each other; however, the "kissing" is never fatal. These fish may be useful as algae eaters to control algae growth. To prevent digging and to present enough surface area for algae growth, the substrate should consist of large-diameter gravel and stones. The aquarium's back glass should not be cleaned during regular maintenance, as the gouramis will feed on the algae grown there. Most plants will not survive the fish's grazing, so inedible plants such as Java fern, Java moss, or plastic plants are recommended.

These fish are omnivorous and need both plant and animal matter in its diet. The fish will accept vegetables such as cooked lettuce and any kind of live food. Water hardness should be between 5-30°dGH and pH between 6.8-8.5; the temperature should be between 22-28°C (72-82°F). When breeding kissing gouramis, soft water is preferred. As the fish do not build a nest, lettuce leaves placed on the water surface serve as a spawning medium. The lettuce will eventually host bacteria and infusoria which the fry will feed upon.

Knifefish

Knifefish refers to several knife-shaped fishes:

- The order Gymnotiformes of knifefishes.
- The family Notopteridae of featherbacks.
- The family Sternopygidae of glass knifefishes.
- Four other fishes not in any of the above:
  - — Grey knifefish, *Bathystethus cultratus*.
  - — Blue knifefish, *Labracoglossa nitida*.
  - — Collared knifefish or Finscale razorfish, *Cymolutes torquatus*.
  - — Jack-knifefish, *Equetus lanceolatus*.

5

Parascylliidae

Parascylliidae is a family of collared carpet sharks in the order Orectolobiformes.

Genera and Species

- Genus *Cirrhoscyllium* (Smith & Radcliffe in Smith , 1913).
  - — Barbelthroat carpetshark, *Cirrhoscyllium expolitum* (Smith & Radcliffe, 1913).
  - — Taiwan saddled carpetshark, *Cirrhoscyllium formosanum* (Teng, 1959).
  - — Saddle carpetshark, *Cirrhoscyllium japonicum* (Kamohara, 1943).
- Genus *Parascyllium* Gill , 1862,
  - — Collared carpetshark, *Parascyllium collare* (Ramsay & Ogilby, 1888).
  - — Rusty carpetshark, *Parascyllium ferrugineum* (McCulloch, 1911).
  - — Ginger carpetshark, *Parascyllium sparsimaculatum* (Goto & Last, 2002).
  - — Necklace carpetshark, *Parascyllium variolatum* (Dumeril, 1853).

Catalufa

Catalufa is the common name for three species of fish belonging to the Priacanthidae family:

- *Heteropriacanthus cruentatus*
- *Priacanthus arenatus*
- *Pristigenys serrula* - the popeye catalufa

The word *catalufa* is also used in several Caribbean countries as the Spanish or French language common name for a number of other Priacanthidae species. The French-speaking islands of Martinique and Guadeloupe also refer to the species *Rhomboplites aurorubens* (family Lutjanidae) as a catalufa.

Catalufas are also commonly called "bigeyes" (the common name for all Priacanthidae).

Catfish

Catfish (order Siluriformes) are a very diverse group of bony fish. Named for their prominent barbels, which give the image of cat-like whiskers. They feature some of the smallest known vertebrates, including the *candiru,* the only vertebrate parasite to attack humans, as well as Mekong giant catfish, the largest reported freshwater fish. There are armour-plated types and also naked types, neither having scales.

Despite their common name, not all catfish have prominent barbels; what defines a fish as being in the order Siluriformes are in fact certain features of the skull and swimbladder. Catfish are of considerable commercial importance; many of the larger species are farmed or fished for food, and some are exploited for sport fishing, including a kind known as noodling. Many of the smaller species, particularly the genus *Corydoras,* are important in the aquarium hobby.

Taxonomy

Catfish belong to a superorder called the Ostariophysi, which also includes the Cypriniformes, Characiformes, Gonorynchiformes and Gymnotiformes, a superorder characterised by the Weberian apparatus. Some place Gymnotiformes as a suborder of Siluriformes, however this is not as widely accepted. Currently, the Siluriformes are said to be the sister group to the Gymnotiformes, though this has been debated due to more recent molecular evidence. As of 2007 there are about 36 extant catfish families, and about 3,023 extant species have been described. This makes the catfish order the second or third most diverse vertebrate order; in fact, 1 out of every 20 vertebrate species is a catfish.

The taxonomy of catfishes is quickly changing. In a 2007 paper, *Horabagrus, Phreatobius,* and *Conorhynchos* were not classified under any current catfish families.

There is disagreement on the family status of certain groups; for example, Nelson (2006) lists Auchenoglanididae and Heteropneustidae as separate families, while the All Catfish Species Inventory (ACSI) includes them under other families. Also, FishBase and the Integrated Taxonomic Information System lists Parakysidae as a separate family, while this group is included under Akysidae by both Nelson (2006) and ACSI. Many sources do not list the recently revised family Anchariidae.

Thus, the actual number of families differs between authors. The species count is in constant flux due to taxonomic work as well as description of new species. On the other hand, our understanding of catfishes should increase in the next few years due to work by the ACSI.

Except for Diplomystidae, the most primitive family of catfish, the relationship between the families is relatively unknown. Classifications of superfamilies varies. Many catfish families are classified into their own superfamilies. Only a few superfamilies contain more than one family. In Nelson (2006) these are Loricarioidea (Amphillidae, Trichomycteridae, Nematogenyidae, Callichthyidae, Scoloplacidae, Astroblepidae, Loricariidae), Sisoroidea (Amblycipitidae, Akysidae, Sisoridae, Erethistidae, Aspredinidae), Doradoidea (Mochokidae, Doradidae, Auchenipteridae), Siluroidea (Siluridae, Malapteruridae, Auchenoglanididae, Chacidae, Plotosidae,

Clariidae, Heteropneustidae), and Bagroidea (Austroglanididae, Claroteidae, Ariidae, Schilbeidae, Pangasiidae, Bagridae, and Pimelodidae). In a recent phylogenetic analysis, however, alternative superfamilies with different constituent families were proposed. In this study, the superfamilies are Clarioidea (Clariidae, Heteropneustidae), Arioidea (Ariidae, Anchariidae), Pimelodoidea (Pimelodidae, Pseudopimelodidae, Heptapteridae, Conorhynchos), Ictaluroidea (Ictaluridae, Cranoglanididae), and Doradoidea (Doradidae, Auchenipteridae). Also, Sisoroidea, unlike that of Nelson, does not include Aspredinidae. The other remaining families are classified by themselves or grouped with other families or superfamilies in larger unranked clades.

The rate of description of new catfishes is at an all-time high. Between 2003 and 2005, over 100 species have been named, a rate three times faster than that of the past century. In June, 2005, researchers named the newest family of catfish, Lacantuniidae, only the third new family of fish distinguished in the last 70 years (others being the coelacanth in 1938 and the megamouth shark in 1983). The new species in Lacantuniidae, *Lacantunia enigmatica*, was found in the Lacantun river in Chiapas, Mexico.

Evolution

A number of catfish fossils are known. Catfishes often have large, heavy bones that lend themselves to fossilisation and, comparatively large otoliths. As such, a large number of species of catfishes have been named from complete or partial skeletal fossils or even from only otoliths. 19 valid genera and 72 species are based exclusively on fossil remains.

The earliest known catfish are known from the late Campanian-early Maastrichtian of Argentina. Fossils of the Eocene period have been found from Seymour Island in Antarctica. The order dispersed early throughout the continents primarily through land bridges. Australian species of catfish are all species from families that can enter saltwater; these fish travelled to Australia this way, and then reverted to a freshwater lifestyle.

The catfish must have spread through Africa to Asia during the late Jurassic if they were to reach Asia. During the Cretaceous period, the rift between South America and Africa would be forming; this may explain the contrast in families between the two continents. Most of the freshwater catfish of the two continents appear to be completely unrelated. Their relatively low diversity in Africa may explain why some primitive fish families coexist with them while they are absent in South America, where the more primitive fish may have been driven extinct. The earliest they could have spread into Central America was the late Miocene.

Distribution and Habitat

Extant catfish species live in inland or coastal waters of every continent except Antarctica. Catfish have inhabited all continents at one time or another. Catfish are most diverse in tropical South America, Africa, and Asia. More than half of all catfish species live in the Americas. They are the only ostariophysans that have entered freshwater habitats in Madagascar, Australia, and New Guinea.

They are found primarily in freshwater environments of all kinds, though most inhabit shallow, running water habitats. Some species even inhabit caves. *Phreatobius cisternarum* is even known to live underground in phreatic habitats. Numerous species from the families Ariidae and Plotosidae, and a few species from among the Aspredinidae and Bagridae, are also found in marine environments.

Ecology

Most catfish are benthic in nature, meaning they normally associate with the bottom of the water column. However, variety of other lifestyles are also represented among the catfishes. A few species are pelagic in nature. The candirú is a parasitic catfish that can attack humans. *Panaque* is a genus of catfishes that are the only fishes able to eat and digest wood.

Catfishes express varying levels of care of reproductive strategies. Internal insemination is probable in all species of

Auchenipteridae. In loricariids, parental care is usually well-developed and the male guards the eggs and sometimes the larvae, either carrying eggs or having the eggs attached to the underside of rocks or in cavities. In most of Ariidae, if not all species, the male is a mouthbrooder; he carries the relatively large eggs in his mouth until the young hatch.

Physical Characteristics

External Anatomy: Most catfish are adapted for a benthic lifestyle. In general, they are negatively buoyant, which means that they will usually sink rather than float due to a reduced gas bladder and a heavy, bony head. Catfish have a variety of body shapes, though most have a cylindrical body with a flattened ventrum to allow for benthic feeding.

A flattened head allows for digging through the substrate as well as perhaps serving as a hydrofoil. Most have a mouth that can expand to a large size and contains no incisiform teeth; catfish generally feed through suction or gulping rather than biting and cutting prey. However, some families, notably Loricariidae and Astroblepidae, have a suckermouth that allows them to fasten themselves to objects in fast-moving water. Catfish also have a maxilla reduced to a support for barbels; this means that they are unable to protrude their mouths as other fish such as carp.

Catfish have no scales; their bodies are either naked or covered in bony plates called scutes. In some species, the mucus-covered skin is used in cutaneous respiration, where the fish breathes through its skin. They may have up to four pairs of barbels: nasal, maxillary (on each side of mouth), and two pairs of chin barbels, although pairs of barbels may be absent, depending on the species.

Because their barbels are more important in detecting food, the eyes on catfish are generally small. Like other ostariophysans, they are characterised by the presence of a Weberian apparatus. Their well-developed Weberian apparatus and reduced gas bladder allow for improved hearing as well as sound production.

All catfish, except members of Malapteruridae (electric catfish), possess a strong, hollow, bonified leading spine-like ray on their dorsal and pectoral fins. As a defence, these spines may be locked into place so that they stick outwards, which can inflict severe wounds.

In several species catfish can use these fin rays to deliver a stinging protein if the fish is irritated. This poison is produced by glandular cells in the epidermal tissue covering the spines. In members of the family Plotosidae, and of the genus *Heteropneustes*, this protein is so strong it may hospitalise humans unfortunate enough to receive a sting; in *Plotosus lineatus*, the stings may result in death.

Size

Catfish have one of the greatest range in size within a single order of bony fish. Catfish range in size and behaviour from the heaviest, the Mekong giant catfish in South East Asia and the longest, the wels catfish of Eurasia, to detritivores (species that eat dead material on the bottom), and even to a tiny parasitic species commonly called the candiru, *Vandellia cirrhosa*. Some of the smallest species of Aspredinidae and Trichomycteridae reach sexual maturity at only 10 millimetres (.4 in). Many catfish have a maximum length of under 12 cm.

The wels catfish, *Silurus glanis*, is the only native catfish species of Europe, besides the much smaller related Aristotle catfish found in Greece. Mythology and literature record wels catfish of astounding proportions, yet to be scientifically proved. The average size of the species is about 1.2 m to 1.6 m, and fish more than 2 m are very rare.

The largest specimens on record measure more than 2.5 m in length and sometimes exceeded 100 kg. The wels catfish was introduced to Britain, Italy, Spain, Greece and some other countries during the last century. The species has flourished in the warm lakes and rivers of Southern Europe. The River Danube, River Po in Italy and the River Ebro in Spain are famous for huge wels catfish, which grow up to 2 m. These habitats contain plenty of food and lack natural predators.

Ictalurus furcatus, in the Mississippi River on May 22, 2005 that weighed 56.25 kg (124 lb). The largest flathead catfish, *Pylodictis olivaris*, ever caught was in Independence, Kansas, weighing 56 kg (123 lb 9 oz).

However, these records pale in comparison to a giant Mekong catfish caught in northern Thailand in May 1, 2005 and reported to the press almost 2 months later, that weighed 293 kg (646 lb). This is the largest giant Mekong catfish caught, but only since Thai officials started keeping records in 1981. The giant Mekong catfish are not well studied since they live in developing countries and it is quite possible that they can grow even larger.

Internal Anatomy

The retina of catfish are composed of single cones and large rods. Many catfish have a tapetum lucidum which may help enhance photon capture and increase low-light sensitivity. Double cones, though absent in most teleosts are absent from catfish.

The anatomical organisation of the testis in catfish is variable among the families of catfish, but the majority of them present fringed testis: Ictaluridae, Claridae, Auchenipteridae, Doradidae, Pimelodidae, and Pseudopimelodidae.

In the testes of some species of Siluriformes, organs and structures such as a spermatogenic cranial region and a secretory caudal region are observed, in addition to the presence of seminal vesicles in the caudal region. The total number of fringes and their length are different in the caudal and cranial portions between species. Fringes of the caudal region may present tubules, in which the lumen is filled by secretion and spermatozoa. Spermatocysts are formed from cytoplasmic extensions of Sertoli cells; the release of spermatozoa is allowed by breaking of the cyst walls.

Fish ovaries may be of two types: gymnovarian or cystovarian. In the first type, the oocytes are released directly into the coelomic cavity and then eliminated. In the second type, the oocytes are conveyed to the exterior through the oviduct. Many catfish are cystovarian in type, including *Pseudoplatystoma corruscans, P. fasciatum, Lophiosilurus alexandri,* and *Loricaria lentiginosa.*

Catfish as Food

Catfish have been widely caught and farmed for food for hundreds of years in Africa, Asia, and Europe. Judgements as to the quality and flavour vary, with some food critics considering catfish as being excellent food, others dismiss them as watery and lacking in flavour. In Central Europe catfish were often viewed as a delicacy to be enjoyed on feast days and holidays. Migrants from Europe and Africa to the United States brought along this tradition, and in the southern United States catfish is an extremely popular food. The most commonly eaten species in the United States are the channel catfish and blue catfish both of which are common in the wild and increasingly widely farmed. Catfish is eaten in a variety of ways, in Europe it is often cooked in similar ways to carp but in the United States it is typically breaded with cornmeal and fried. In Indonesia catfish are very popular food. They are usually served grilled in street stalls called warung and eaten with vegetables, the dish is called *Pecel Lele* (*Lele* is the Indonesian word for catfish). The iridescent shark is a common food fish in parts of Asia.

Catfish is also high in Vitamin D.

Aquaculture

Catfish are easy to farm in warm climates, leading to inexpensive and safe food in local grocers. Ictalurids are cultivated in North America (especially in the Deep South, with Mississippi being the largest domestic catfish producer). Channel catfish (*Ictalurus punctatus*) supports a $450 million/ yr to aquaculture industry.

In Asia, many catfish species are important as food. Several walking catfish (Clariidae) and shark catfish (Pangasiidae) species are heavily cultured in Africa and Asia. Exports of one particular shark catfish species from Vietnam, *Pangasius bocourti,* has met with pressures from the US catfish industry. In 2003, The United States Congress passed a law preventing the imported fish from being labelled as catfish. As a result, the Vietnamese exporters of this fish now label their products sold in the US as "basa fish."

There is a large and growing ornamental fish trade, with hundreds of species of catfish, such as *Corydoras* and armoured suckermouth catfish (often called plecos), being a popular component of many aquaria. Other catfish commonly found in the aquarium trade are banjo catfish, talking catfish, and long-whiskered catfish.

Catfish as Invasive Species

Representatives of the genus *Ictalurus* have been misguidedly introduced into European waters in the hope of obtaining a sporting and a food resource. However, the European stock of American catfishes has not achieved the dimensions of these fishes in their native waters, and have only increased the ecological pressure on native European fauna. Walking catfish has also been introduced in the freshwaters of Florida, with the voracious catfish becoming a major alien pest there. Flathead catfish, *Pylodictis olivaris,* is also a North American pest on Atlantic slope drainages. Armoured suckermouth catfish, released by aquarium fishkeepers, has also established feral populations in many warm waters around the world.

Catla

Gibelion catla, the only member of the genus *Gibelion,* of the carp family Cyprinidae is a fish with a large protruding lower jaw. It is commonly found in rivers and freshwater lakes in the South and South East Asia.

Catshark

The cat sharks or catsharks are a family (Scyliorhinidae) of sharks, with over 110 species recorded. Paradoxically perhaps, while the group is called the cat shark family, many species are commonly called dogfish.

Cat sharks may be distinguished by their elongated cat-like eyes and two small dorsal fins set far back. Most species are not particularly large, with lengths up to 60-70 cm or so, although the humpback cat shark, *Apristurus gibbosus,* from the deep waters of the South China Sea has been recorded at 4 m in length.

Most of the species have a patterned appearance, ranging from stripes to patches to spots. They feed on invertebrates and smaller fish. Some species are ovoviviparous while most lay eggs in tough egg-cases with curly tendrils at each end known as "mermaid's purses".

The "swell sharks" of the genus *Cephaloscyllium* have the curious ability to fill their stomachs with water or air when threatened, increasing their girth by a factor of 2 or 3.

The Australian marbled catshark, *Atelomycterus macleayi,* is a favoured type for home aquaria, because it rarely grows to more than 60 cm (2 ft) in length.

Genera

- *Apristurus*
- *Asymbolus*
- *Atelomycterus*
- *Aulohalaelurus*
- *Cephaloscyllium*
- *Cephalurus*
- *Galeus*
- *Halaelurus*
- *Haploblepharus*
- *Holohalaelurus*
- *Parmaturus*
- *Pentanchus*
- *Poroderma*

- *Schroederichthys*
- *Scyliorhinus*

Cladogram

- Schyliorhinidae
  - — Caninoa
  - — Schyliorhininae
  - — Galeinae
    - Pentanchini
    - Galeini
    - Galeina
    - Halelaelurina
  - — Atelomycterininae
  - — Schroedericthyinae

Cavefish

The cavefishes (commonly: *blindfishes, swampfishes*) are a family (Amblyopsidae) of fish found in caves and adapted to life in the dark, notably lacking eyes and pigmentation, as a result having a pale or whitish colour. They are all found in the southern and eastern United States. Some can be found elsewhere as well. There are about 80 known varieties of cavefish.

Cavefishes are generally small, ranging up to 11 cm in length. Most do not have pelvic fins, although *Amblyopsis spelaea* has small ones with up to six rays. Only three species of cave fish lack eyes completely, but several others have useless eyes. The majority of cave fish have little to no pigment in their skin. These features are an example of regressive evolution.

Cavefishes can only be found in caves that have streams running *into* them; a cave with no inlets (such as Blanchard Springs Caverns in Arkansas) will not contain cavefishes. They are believed to have been evolved from their aboveground counterparts.

Species

The family includes six species in five genera:

- Genus *Amblyopsis Genus* Typhlichthys
  - Southern cavefish, *Typhlichlhys subterraneus* (Girard, 1859).

Chain Pickerel

The chain pickerel, *Esox niger* (syn. *Esox reticulatus*), is a species of freshwater fish in the pike family (family Esocidae) of order Esociformes. It is also known as the federation pike or federation pickerel. Its range is along the eastern coast of North America from southern Canada to Florida, and west to Texas.

The chain pickerel has a distinctive chainlike pattern on its sides and its body resembles that of the northern pike. It typically reaches 24 inches in length with a weight of 3-5 pounds. The US record is over nine pounds.

The chain pickerel feeds primarily on smaller fish which it ambushes from cover with rapid lunges and secures with its sharp teeth.

It is a popular sport fish. It is an energetic fighter on the line. Anglers after pickerel have success with live minnows, spinnerbaits, spoon lures, and other lures. Practically every bass lure can be effective for pickerel, although they seem to be particularly susceptible to flashy lures, which imitate small prey fish. Dragging a plastic worm, lizard, frog, and other soft plastics can also be extremely effective.

In ponds and smaller lakes however, some anglers see pickerel as a threat to trout populations and trout restocking efforts. It is sometimes considered an easy-to-bag "trash fish", and not particularly tasty. A commonly used nickname in the southeast for this fish is the Southern Pike.

This fish is edible ,but there is a special way to clean them to get the meat with no bone. You must first clean them as any other fish, then carefully cut the backbone out. Then you fillet them in three to four inch segments .You then carefully cut small ,preferably 1cm, incisions on the meat part of the fish up the flesh and sideways. Then deepfry the fillets until brown. No bone will be in it.

Red Drum

The red drum (*Sciaenops ocellatus*), also known as channel bass, redfish, puppy drum or just red, is a game fish that is found in the Atlantic Ocean from Massachusetts to Florida and in the Gulf of Mexico from Florida to Northern Mexico. It is the only species in the genus *Sciaenops.*

Characteristics

Red Drum usually occur in great supply along the Tar Heel coastal waters and have been found to weigh up to 94 pounds although most large ones average between 30 and 40 pounds. The all-tackle world record of 94 pounds 2 oz. was set in Nov. 7, 1984 near Avon, NC. The previous world record of 90 pounds, was set in nearby Rodanthe, NC, 11 years prior, to the day, on Nov. 7, 1973. Edible specimens range from 4 to 9 pounds in size. Smaller fish are usually protected by regulation and require release. Larger fish typically are coarser and less tasty, and their sporting qualities tend to cause sportsmen to release them to be caught again.

Mature Red Drum spawn in open oceans, bays and inlets. Juvenille red drum typically inhabit inland lakes and coastal marshes until they reach maturity between 3 and 6 years of age.

Relationship to Humans

The North Carolina General Assembly of 1971 designated the Red Drum as the official State Salt Water Fish.

Red Drum are notable game fish, appealing to fly fishermen and anglers with spinning and baitcasting tackle alike. In shallow water settings, Red Drum are often seen "tailing," or grubbing for food in such a manner that their tail fins are exposed above the surface of the water.

Sight fishing for tailing Red Drum is an increasingly popular past time all along the United States' East and Gulf Coasts.

The Greatest concentrations of Red Drum occur on the Louisiana Gulf Coast, and in particular in the coastal marshes of the Mississippi Delta.

The famous Mosquito Lagoon in Florida is also considered one of the most consistent year round fisheries for large Redfish. Large Reds can also be caught from beaches and piers during the winter months from Jacksonville, FL down to Ponce Inlet, FL.

It has been known to be considered a magical fish that has the powers to heal. Early Carolina settlers would grind the left side of the fish and rub it along their cheeks and neck to heal a sore throat or cough.

On the west coast of Florida, Pine Island Sound and Tarpon Bay, on the inland side of Sanibel Island, are recognised as the premier fisheries for "redfish". The pass between Sanibel and Captiva Islands is even named "Redfish Pass." On the Texas Gulf Coast, a minor bay near the city of Corpus Christi is named "Redfish Bay" and is a popular shallow water destination for anglers seeking the fish.

Once considered poor table fare, the Red Drum gained popularity as a food fish in the early 80's, due largely in part to the Cajun speciality dish "Blackened Redfish" developed by Chef Paul Prudhomme. Red Drum were harvested in great numbers until stocks were dangerously low. Various state and federal wildlife agencies, and conservation groups began a concerted effort to restore Red Drum populations. Red Drum were awarded sports fish status in almost all coastal states and harvests were regulated. Red Drum numbers have since recovered, making them one of the most targeted saltwater game fish in the US.

Channel Catfish

Channel catfish, *Ictalurus punctatus*, are North America's most numerous catfish subspecies. They are also the most fished types of catfish, with approximately 8 million anglers in the USA targeting them per year. A member of the *Ictalurus* genus of American catfishes, channel catfish have a top-end size of approximately 40-50 pounds (18-23 kg).

The world record channel catfish weighed 58 pounds and was caught in 1964 in the Lake Marion, South Carolina. Realistically, a channel catfish over 20 pounds (9 kg) is a spectacular specimen, and most catfish anglers view a 10 pound (4.5 kg) fish as a very admirable catch. Furthermore the average size channel catfish an angler could expect to find in most waterways would be between 2 and 4 pounds. Channel catfish flesh is prized by many anglers and the popularity of channel catfish for food has allowed the rapid growth of aquaculture of this species throughout the United States.

Channel catfish are well distributed throughout the United States and thrive in small rivers, large rivers, reservoirs, natural lakes, and ponds. Channel catfish are omnivores who can be caught on a variety of natural and prepared baits including crickets, nightcrawlers, minnows, shad, chicken livers, frogs, bullheads, sunfish, and suckers. Catfish have even been known to take Ivory Soap as bait .

Channel catfish possess very keen senses of smell and taste. At the pits of their nostrils (nares) are very sensitive odour sensing organs with a very high concentration of olfactory receptors. In channel catfish these organs are sensitive enough to detect several amino acids at about 1 part per 100 million in water. In addition channel catfish have taste buds distributed over the surface of their entire body. These taste buds are especially concentrated on the channel catfish's 4 pairs of barbels (whiskers) surrounding the mouth—about 25 buds per square millimetre. This combination of exceptional senses of taste and smell allows the channel catfish to find food in dark, stained, or muddy water with relative ease.

This combined with the fact that channel catfish will readily scavenge for food explains why cutbaits (fresh cut pieces of fish—usually minnows, shad, herring, sunfish, suckers, etc.) are particularly effective for catching this species of catfish. In addition prepared baits such as dipbaits, punchbaits, bloodbaits, and other "stinkbaits" can be effective in many situations. These baits usually are made from some combination of ground fish, chicken, beef, cheese, sour grains, and many other "secret" ingredients.

Catfish trapping is regulated in some states. Catfish traps include "slat traps," long wooden traps with an angled entranced, and wire hoop traps. Typical bait for these traps include rotten cheese and dog food. Catches of as many as 100+ fish a day are common in catfish traps.

Catfish trapping, however, has recently come under media scrutiny due to the recovery process. The inherent nature of a trap means that the fish can be confined to small areas for, at times, up to twenty-four hours before traps are checked. The channel catfish requires a full range of motion in order to perform aerobic respiration, but since this is not possible in many traps the catfish suffocate. Animal rights activists believe that federal regulations for larger trap sizes should be put in place.

Seema

Seema, also sima and sema, Japanese salmon, Masu salmon, or cherry salmon, *Oncorhynchus masu*, is a salmon of the western Pacific Ocean: (Kamchatka, Kuril Islands, Sakhalin, Primorsky Krai, Korea, Japan). A landlocked subspecies commonly called the Taiwanese salmon or Formosan salmon (*Oncorhynchus masou formosanum*) also exists in Taiwan. This fish prefers a temperate climate, around the area of 65° N - 58° N, and in the sea, it prefers a depth range of 0-200 m.

Appearance

A seema which has reached sexual maturity has a darkened back, and the stripes on the body sides become bright red with crimson tinge to merge on the abdomen into one common longitudinal band of lighter colour. It is for this reason that it was given the name Cherry Salmon.

As adults, seema tend to weigh 2 to 2.5 kg and measure roughly 50 cm in length.

The maximum size that can be attained by this species (which is in the region of Primorsky Krai) is 71 cm long and 9 kg in weight.

Life Cycle

Like other Pacific salmon, its life cycle is subdivided into marine and freshwater periods; in rivers, this species lives from 1 to 3 years and can form living freshwater forms. The sea life cycle, depending on the age of the young, continues for 2 to 3.5 years.

In the sea, the seema feeds intensely on crustaceans, less often on young fish. On attaining sexual maturity, in its third to seventh years of life it enters rivers to spawn. Its spawning run starts earlier than that of other salmon species.

After spawning, most passing fish die, and those that remain alive (preferentially dwarf males) participate in spawning next year, too. Emerging from the nest, the young do not roll into the sea but remain in spawning areas, in the upper reaches of rivers, and on shallows with weak currents. The young move to pools and rolls of the river core to feed on chironomid, stone fly and may fly larvae, and on air insects. The seema rolls into the sea in its second, occasionally even third year of life.

Economic Importance

This salmon, like most others, is a highly commercial species caught in fisheries, raised for aquaculture, and sought after as a game fish. It is marketed fresh and frozen and is often eaten broiled or baked.

Bibliography

Anderson, L. and Bryant, M. D.: *Fish Passage at Road Crossings,* Pacific Northwest Forest and Range Experimental Station, Portland, 1980.

Arnal, R. E.: *Limnology, Sedimentation and Microorganisms of the Salton Sea,* Geological Society of America Bulletin, California, 1961.

Aronson, L. R. and Kaplan, H.: *Function of the Teleostean Forebrain,* University of Chicago Press, Chicago, 1968.

Axelrod, Herbert R. and Warren E. Burgess. *Saltwater Aquarium Fish,* Neptune City, T. F. H. Publications, New Jersey, 1973.

Ayling, T. and Geoffrey J: *Guide to the Sea Fishes,* William Collins Publishers, Auckland, 1987.

Bainbridge, R.: *Problems of fish,* Symposium of the Zoological Society, London, 1961.

Baker, C.O. and Votapka, F.E. *Fish Passage Through Culverts,* Forest Service Technology and Development Center, San Dimas, 1990.

Balfour, F. M.: *A Monograph on the Development of Elasmobranch Fishes,* MacMillan, London, 1878,

Ballinger, C. A. and Drake P. G.: *Culvert Repair Practices Manual,* Turner Fairbank Highway Research Center, Virginia, 1995.

Bardach, J. E. and Villars, T.: *The Chemical Senses of Fishes,* Academic Press, London, 1974.

Bartos, L. R.: *Peak Flow Hydrology in Relation to Bridge and Culvert Design Problems in Southeast Alaska,* National Applied Wildland Hydrology Workshop, Georgia, 1978.

Bath, H.: *Check List of the Fishes of the Eastern Tropical Atlantic,* UNESCO, Paris, 1990.

Bell, M.: *Fisheries Handbook of Engineering Requirements and Biological Criteria,* North Pacific Division, Portland, 1990.

Beschta, R. L.: *Road Drainage Structures,* Culvert Sizing at Stream Crossings, Oregon State University, Oregon, 1984.

Bohlke, E. B.: *Fishes of the Western North Atlantic,* Yale University, New Haven, 1989.

Breder, C. M. and Rosen D. E.: *Modes of Reproduction in Fishes,* Natural History Press, New York, 1966,

Chaplin, C. C. G.: *Fishes of the Bahamas and Adjacent Tropical Waters,* University of Texas Press, Austin, 1993.

Cohen, D. M.: *Fishes of the North Eastern Atlantic and the Mediterranean,* UNESCO, Paris, 1986.

Collette, B. B.: *Check List of the Fishes of the Eastern Tropical Atlantic,* UNESCO, Paris, 1990.

Compagno, L. J. V.: *Sharks of the Order Carcharhiniformes,* Princeton University Press, Princeton, 1988.

Dane, B. G.: *Culvert Guidelines: Recommendations for the Design and Installation of Culverts in British Columbia to Avoid Conflict with Anadromous Fish,* Fisheries and Marine Service, Columbia, 1978.

Dawson, C. E.: *Fishes of the Western North Atlantic,* Yale University, New Haven, 1982.

Dean, B.: *Fishes, Living and Fossil,* MacMillan, New York, 1895.

Denton, E. J., Gray J. A. B. and Blaxter J. H. S.: *The Mechanics of the Clupeid Acoustico Lateralis System,* United Kingdom, 1979.

Derksen, A. J.: *Evaluation of Fish Passage Through Culverts at the Goose Creek Road Crossing near Churchill,* Manitoba Department of Natural Resources, Manitoba, 1977.

Dimeo, A.: *Correcting Vertical Fish Barriers: Investigation of Steeppass and Fish Ladders*, Equipment Development Center, Montana, 1977.

Engle, P.: *Fish Passage Facilities for Culverts of the MacKenzie Highway*, Canada Centre for Inland Waters, Burlington, 1974.

Eschmeyer, W. N., Mead G. W. and Parr A. E.: *Fishes of the Western North Atlanti*, Yale University, New Haven, 1977.

Fichter, George S. and Edward C. Migdalski: *The Fresh and Saltwater Fishes of the World*, Greenwich House, New York, 1983.

Fitch, G. M. *Nonanadromous Fish Passage in Highway Culverts*, Virginia Departent of Transportation, Charlottesville, 1995.

Francis, Malcolm P: *Checklist of the Coastal Fishes of Lord Howe*, Pacific Science, Islands, 1993.

Geistdoerfer, P.: *Check List of the Fishes of the Eastern Tropical Atlantic*, UNESCO, Paris, 1990.

Greenwood, P. H., Miles R. S. and Patterson C.: *Interrelationships of Fishes*, Academic Press, London, 1973.

——————— : *The Cichlid Fishes of Lake Nabugabo*, British Museum of Natural History Bulletin, Uganda,1965.

Haedrich, R. L.: *Check List of the Fishes of the Eastern Tropical Atlantic*, UNESCO, Paris, 1990.

Hanna, G. Dallas.: *Expedition to Guadalupe Island*, California Acad, Mexico,1925.

Hanson, J. A.: *Tolerance of High Salinity by the Pileworm*, California Fish and Game, California, 1972.

Harmelin, Vivien M. L. and Quero J. C.: *Fishes of the North Eastern Atlantic and the Mediterranean*, UNESCO, Paris, 1986.

Hasler, A. D.: *Orientation and Fish Migration*, Academic Press, New York, 1971.

Hauser, Hillary. *Book of Marine Fishes*, Tetra Press, New York, 1984.

Hedderly, Edwin L.: *Fish and Game Conditions in Southern California*, California Fish Game, California 1916.

Heiner, B. and Klavas, P.: *Fish Passage Design at Road Culverts*, Washington Department of Fish and Wildlife, Washington, 1999.

Henderson, P. B.: *Case of Poisoning by the Bonito: The History of Herodotus*, Herodotus, London, 1910.

Herald, Earl Stannard: *Fishes of North America*, Doubleday, New York, 1972.

Hildebrand, Samuel F. and Schroeder, William C.: *Fishes of Chesapeake Bay*, U.S. Bur. Fish, Bull, 1928.

Hoar, W. S. and Randall, D. J.: *Fish Physiology*, Academic Press, New York, 1971.

Holder, Charles Frederick, and Jordan, David Starr: *Fish Stories Alleged and Experienced with a Little History*, Natural and Unnatural, New York,1909.

Hornaday, William J.: *The American Natural History*, New York, 1914.

Hulley, P. A.: *Check List of the Fishes of the Eastern Tropical Atlantic*, UNESCO, Paris, 1990.

Humann, Paul: *Reef Fish Identification*, New World Publications, Orlando, 1992.

Iwamoto T. and Marshall, N. B.: *Fishes of the Western North Atlantic*, Yale University, New Haven, 1973.

Jessop, C. S. and Dryden, R. L.: *Impact Analysis of the Dempster Highway Culvert on the Physical Environment and Fish Resources of Frog Creek*, Fisheries and Marine Service, Canada,1974.

Johnson, A. and Orsborn, J. F.: *Welcome to Culvert College*, Washington Trout, Washington, 1996.

Johnson, James Edward: *Protected Fishes of the United States and Canada*, American Fisheries Society, Maryland: 1987.

Johnson, M. W. *The Copepod Cyclops Dimorphus from the Salton Sea*. American Midland Naturalist, 1953.

Jordan, D.S. and Evermann B.W.: *The Fishes of North and Middle America*, Government Printing Office, Washington, 1900.

Jordan, David Starr: *The Genera of Fishes and a Classification of Fishes,* Stanford University Press, Stanford, 1983.

Jordan, M. C. and Carlson, R. F.: *Design of Depressed Invert Culverts,* University of Alaska, Alaska, 1987.

Joseph, James, Witold Klawe and Pat Murphy: *Tuna and Billfish Fish Without a Country,* Inter American Tropical Tuna Commission, California, 1988.

Kane, D. and McLean R. F.: *Economic Culvert Design Using Fish Swimming Energy and Power Capabilities,* American Fisheries Society, Bates, 1993.

Kane, D. and Wellen, P. M.: *Appendix to: A Hydraulic Evaluation of Fish Passage Through Roadway Culverts in Alaska,* Water Center, Alaska, 1985.

Karrer, C. and Post A.: *Check List of the Fishes of the Eastern Tropical Atlantic,* UNESCO, Paris, 1990.

Katopodis, C. Robinson P. R. and Sutherland, B. G. *A Study of Model and Prototype Culvert Baffling for Fish Passage,* Department of Fisheries and the Environment, Manitoba, 1978.

Leedy, D. L.: *Highway Wildlife Relationships,* Urban Wildlife Research Center, Maryland, 1975.

Lindsey C. C., Robins C. R. and Scott W. B.: *A List of Common and Scientific Names of Fishes,* American Fisheries Society, Washington, 1970.

Linsley, R. H. and Carpelan L. H.: *Invertebrate Fauna,* California Fish and Game, California, 1961.

—————— : *The Pile Worm, Neanthes Succinea Frey and Leukart,* California Fish and Game, California, 1961.

Lowman, B.: *Investigation of Fish Passage Problems Through Culverts,* Equipment Development Center, Montana, 1974.

Maisey, J. G.: *An Evaluation of Jaw Suspension in Sharks,* American Museum of Natural History Novitates, 1980,

Marshall, N. B.: *The Life of Fishes,* Universe Books, New York, 1973.

Matsui, T. and Rosenblatt R. H.: *Review of the Deep-sea Fish Family Platytroctidae,* Oceanogr University, California, 1987.

Mayer, G. F.: *Check List of the Fishes of the Eastern Tropical Atlantic*, UNESCO, Paris, 1990.

McClellan, T.: *Fish Passage Through Highway Culverts*, Federal Highway Administration, Portland, 1971.

McEachran, J. D. and L. V. Compagno: *Interrelationships of and within Breviraja Based on Anatomical Structures*, Bull, Marine, 1982.

Mead, G. M., Olsen Y. H., Breder C. M. and Schroeder W. C.: *Fishes of the Western North Atlantic*, Yale University, New Haven, 1966.

Metsker, H. E.: *Fish Versus Culverts*, Technical Information Center, Washington, 1970.

Miles, R. S., Patterson, C., Rosen, D. E. and Weitzman, S. H.: *Phyletic Studies of Teleostean Fishes, with a Provisional Classification of Living Forms*, Bulletin of the American Museum of Natural History, 1966.

Miller, Daniel J. and Robert N. Lea. *Guide to the Coastal Marine Fishes of*, California Department of Fish and Game, California, 1972.

Monte, Le and Francesca Raimonde: *Giant Fishes of the Open Sea*, Rinehart and Winston, New York, 1985.

Moseley, Charles J.: *The Official World Wildlife Fund Guide to Endangered Species*, Beacham Publishing, Washington, 1992.

Moyle, Peter B. and Joseph J. Cech: *Fishes. An Introduction to Ichthyology*, Prentice Hall, New Jersey,1988.

Nelson, J. S.: *Fishes of the World*, John Wiley and Sons, New York, 1984.

Nielsen, J. C.: *Check List of the Fishes of the Eastern Tropical Atlantic*, UNESCO, Paris, 1990.

Nikolsky, G. V.: *The Ecology of Fishes*, T. F. H. Publications, New Jersey, 1978.

Ommanney, F. D.: *The Fishes*, Time, New York, 1984.

Ono, R. Dana, James D. Williams, and Anne Wagner: *Vanishing Fishes of North America*, Stone Wall Press, Washington, 1983.

Palmer, G.: *Fishes of the North Eastern Atlantic and the Mediterranean,* UNESCO, Paris, 1986.

Parin, N. V.: *Fishes of the North Eastern Atlantic and the Mediterranean,* UNESCO, Paris, 1986.

Parker, R. A. and LeBlanc J. U.: *Northwestern Gulf of Mexico Topographic Features Study,* Texas A&M University, New Orleans, 1978.

Paulin, C. D.: *Papers on the Systematics of Gadiform Fishes,* Natural Hist, Los Angeles, 1989.

Quero, J. C. and Gayet M.: *Check List of the Fishes of the Eastern Tropical Atlantic,* UNESCO, Paris, 1990.

Quero, J. C. and Markle D.F.: *Fishes of the North Eastern Atlantic and the Mediterranean,* UNESCO, Paris, 1984.

Quinn, John R.: *Our Native Fishes,* Countryman Press, Vermont, 1990.

Raimondi, P. T.: *Adult Plasticity and Rapid Larval Evolution in a Recently Isolated Barnacle Population,* Biological Bulletin, California, 1992.

Randall, D. J.: *Fish Physiology,* Academic Press, London, 1970.

Robins, C. R., Bailey R. M., Bond C. E. and ScottW. B.: *Common and Scientific Names of Fishes from the United States and Canada,* American Fisheries Society, Maryland, 1980.

Roux, C.: *Check List of the Fishes of the Eastern Tropical Atlantic,* UNESCO, Paris,1990.

Sakurai, Atzushi, Yohei Sakamoto and Fumitoshi Mori: *Aquarium Fish of the World,* Chronicle Books, San Francisco, 1991.

Slatick, E.: *Passage of Adult Salmon and Trout Through Pipes,* US Fish and Wildlife Service, Washington, 1970.

Smith, C. L.: *Check List of the Fishes of the Eastern Tropical Atlantic,* UNESCO, Paris, 1990.

Smith, D. G.: *Check List of the Fishes of the Eastern Tropical Atlantic,* UNESCO, Paris, 1990.

——————— : *Fishes of the Western North Atlantic,* Yale University, New Haven, 1989.

Smith, Vaniz, W. F. *Fishes of the North Eastern Atlantic and the Mediterranean*, UNESCO, Paris, 1986.

Sulak, K. J.: *Fishes of the North Eastern Atlantic and the Mediterranean*, UNESCO, Paris, 1984.

Tee, Van, J., Breder C. M., Hildebrand S. F. and Parr A. E.: *Fishes of the Western North Atlantic*, Yale University, New Haven, 1948.

Thomas, J. A. and Miles R. S.: *Paleozoic Fishes*, Chapman and Hall, London, 1971,

Thompson, Peter: *Thompson's Guide to Freshwater Fishes*, Houghton Mifflin, Boston, 1985.

Thomsen, Marie Louise: *The Home of the Fish*: Journal of Cuneiform Studies California, 1975.

Thresher, R. E.: *Reproduction in Reef Fishes*, T. F. H. Publications, New Jersey, 1984.

Travis, M. D. and McLean, R. F.: *Fundamentais of Culvert Design for Passage of Weak Swimming Fish*, Fairbanks, Alaska,1991.

Travis, M. D. and Tilsworth, T.: *Fish Passage Through Poplar Grove Creek Culvert, Transportation Research* Alaska, 1986.

Wheeler, Alwyne: *Fishes of the World*, MacMillan Publishing, New York, 1975.

Whitehead, P. J. P. and Bauchot M. L.: *Fishes of the North Eastern Atlantic and the Mediterranean*, United Nations Educational, Scientific and Cultural Organization, Paris, 1986.

Wilson, Josleen: *North American Fish*, Gramercy Books, New York, 1991.

Woods, L. P. and Sonoda P. M.: *Fishes of the Western North Atlantic*, Yale University, New Haven, 1973.

Wootton, R. J.: *Ecology of Teleost Fishes*, Chapman & Hall, New York, 1990.

Index

D

H

T

W

❑❑❑

R

S